Lecture Notes in Biomathematics

Managing Editor: S. Levin

21

Theoretical Approaches to Complex Systems

Proceedings, Tübingen, June 11–12, 1977

Edited by
R. Heim and G. Palm

Springer-Verlag
Berlin Heidelberg New York 1978

Editors
Roland Heim
Institut für Informationsverarbeitung
Universität Tübingen
Köstlinstraße 6
7400 Tübingen/BRD

Günther Palm
Max-Planck-Institut
für Biologische Kybernetik
Spemannstraße 38
7400 Tübingen/BRD

AMS Subject Classifications (1970): 92-02, 92A05, 93C10, 93DXX

ISBN 3-540-08757-5 Springer-Verlag Berlin Heidelberg New York
ISBN 0-387-08757-5 Springer-Verlag New York Heidelberg Berlin

Printing and binding: Beltz Offsetdruck, Hemsbach/Bergstr.
2141/3140-543210

Preface

This volume is the issue of the symposium 'Theoretical Approaches to Complex Systems', held in Tübingen on June 11 and 12, 1977. The meeting was dedicated to the late Ernst Pfaffelhuber of the Institute for Information Sciences, University of Tübingen, in memory of his contributions to information theory, physics, and biology and of his many fruitful contacts with the Tübingen scientists gathered on this occasion.

We would like to express our gratitude to all participants for their willingness to offer a weekend, for their help and the care they took in writing a contribution to this volume.

R. Heim
G. Palm

CONTENTS

Introduction

When a 20th century scientist is concerned with some phenomena, it usually turns out that the 'system' he is studying is very 'complex'.

It has not always been like that. Looking back we have the feeling that formerly - perhaps in the 18th century - it seemed that the physical phenomena were understood perfectly. Later the Hamiltonian formalism allowed a rigorous analysis of a large class of phenomena (e.g. the movements of planets).

This contrast may be a consequence of an enlargement of the class of 'physical phenomena' since that time, which occured in two ways:

1. To account for minute details observed in experiments of increasing accuracy which could not be explained by classical mechanics, new and more complicated theories were developed, e.g. quantum mechanics and relativity theory. New fundamental forces, interactions, and particles were postulated and detected.
2. The class of phenomena investigated was extended to those which were known in the 18th century but either deliberately excluded from the physical theory (e.g. turbulence and phase transitions) or even considered to be inaccessible to rigorous theories (e.g. all biological phenomena).

As a consequence of this development, today there is no unified theory that can accurately describe all known physical phenomena. Instead, the point of view has been adopted that one should build special mathematical models depending on the class of phenomena investigated.

This volume presents various views on systems whose complexity arises from the kind of observations mentioned in 2. In particular, most of the contributions are concerned with biological phenomena.

In the analysis of such systems, one way to proceed is to connect a phenomenological description with comparatively simple underlying principles, e.g. the phenomenological theory of heat to classical mechanics or the behavior of an animal to rules governing neuronal interactions. In these cases we have a large number of components of identical type or of only a few types, and the problem is to describe and predict the macroscopic behavior of the whole system by knowing the functioning of the microscopic components and the rules governing their interactions. This kind of complexity we may call <u>cooperative complexity</u>.

Another way to understand biological systems is to ask what purpose they are designed for. Norbert Wiener stressed the importance of this idea for a fruitful cooperation between engineers and biologists. Implicitely, these considerations on design had been used in biology before. For example, the purely 'biological' description of the human eye involves the term 'lens' which certainly stems from engineering.

One has to be careful in applying this idea, since the building blocks available to the engineer are often quite different from those available to nature (e.g. transistors vs. neurons). One should try to find a simple design for the organization of a certain behavior in terms of neurons instead of transistors, although a technical realization would be exceedingly complicated. Today, the behavior of such a design can be predicted by computer simulation. For a deeper insight into the corresponding mathematical model (e.g. differential equations) as well as for the problem of cooperative complexity new mathematical tools are necessary.

In this book some mathematical tools are presented which may be helpful in analyzing complex systems (H.Hahn, G.Palm, O.Rössler and P.J.Ortoleva). Some concrete biological systems exhibiting cooperative complexity are discussed, introducing interesting mathematical approaches (W.U.an der Heiden, T.Poggio and V.Torre, W.Reichardt) or considerations on design (M.Conrad, V.Braitenberg, D. Varjú). Some again more mathematical papers employ considerations on design in order to develop optimality criteria, which may be applied also to biological systems (R.Heim, M.Dal Cin).

G. Palm

R. Heim

Existence Criteria for Bifurcations as Stability Criteria for Critical Nonlinear Control Systems

Hubert Hahn

Fachbereich Physik,Universität Tübingen,Federal Republic of Germany

INTRODUCTION

Critical systems are defined as systems whose stability behavior cannot be decided in first approximation.The linear part of the corresponding operator has at least one eigenvalue with vanishing real part and the rest with negative real parts.
From the pure mathematical point of view critical systems may be regarded as exceptional or even as pathological cases.From the point of view of physics (e.g. analytical mechanics,compare Hamiltonian Systems) these systems are already far away from being exceptional cases.In nonlinear control-technique as well as in nonlinear network-technique critical nonlinear systems act as a main source and at the same time as an organizing principle of a variety of different qualitative phenomena which in turn are among the basic ingredients of intuitive engineering thinking in the laboratory.The ability of a nonlinear dynamical system to produce phenomena like synchronization,triggering,mono- or bistability ,relaxation oscillations and switching- or storage-behavior is intimately connected to its critical behavior.Unfortunately critical nonlinear systems are -up to some special exceptional approaches- basically excluded from most of the wellknown theoretical approaches in network- and control-theory.This situation may be characterized in terms of the following (somewhat overrefined) statement:

> The design-engineer thinks in terms of singular (qualitative) concepts, but he computes on the basis of regular (quantitative) theories.

This is -according to the authors mind- one of the main reasons for the often called "gap between theory and practice" in control and net-worktechnique (H.Hahn,1975).Moreover,there are good reasons to expect, that the development of theories which incorporate the critical behavior of nonlinear systems as basic ingredients,will correct a wide-spread opinion concerning the principal differences in the behavior of technical (physical) and biological systems.
The most comprehensive (but still not yet complete) investigations with respect to the stability of critical nonlinear systems are due to Malkin and Liapunov (compare W.Hahn,1967,W.Reissig et al.,1969).

Further research in this direction has been done by Bautin,who has shown that the stability behavior of a nonlinear system near the stability-boundary is determined by its stability behavior on the boundary(N.N. Bautin,195o; W.Hahn,1967).There are many practical arguments for a systematic study and a deeper understanding of the stability behavior of nonlinear control-systems near the stability boundary resp.at the limits of the parameter-space.One rather actual reason may be found in recent investigations concerning the "Limits to Growth" in the economical world (W.Forrester,1971;D.Meadows,1972;M.Mesarovic,E.Pestel, 1972).

In a recent paper,special aspects of the problem of practical stability have been treated from the point of view of bifurcation-theory (H.Hahn, 1976).A critical system behavior turns out to be a necessary condition for the phenomenon of bifurcation (compare J.Kelly,1967,J.Hale,1971,M. Hausrath,1973,J.E.Marsden and M.Cracken,1976).

In this paper,the concepts of a "branch-point"of a system,the "smooth embedding" of a stationary solution and the "parametrically perturbed related system" (related to a given critical system) are introduced to investigate the stability of a critical nonlinear system. In order to clarify the basic idea of this approach only the simplest case of a critical system (there exists exactly one simple critical eigenvalue) is treated.More general critical systems will be treated in further investigations using the same ideas.

Applying Malkin's stability results and a special technique from bifurcation-theory (known as "Newton's Diagram") statements are proved, which guarantee the equivalence between the two concepts of the asymptotic stability of a critical system and the existence of special pattern of stationary solutions which branch off the trivial solution of a "parametrically perturbed related system".As a consequence,the stability behavior of a critical nonlinear system can be decided from the knowledge of the pattern of the corresponding branching solutions and from linear eigenvalue theory.Consequently these results provide a <u>purely algebraic</u> procedure to decide the stability behavior of critical nonlinear systems.Moreover these criteria have strong appeal to <u>geometry</u> and <u>can be verified in experiments</u>.This is not only of importance from the point of view of applications,but it provides also a deeper understanding of the underlying phenomena.

A generalization of the theorems (incorporating multiple critical eigenvalues as well as complex conjugate eigenvalues with vanishing real parts) is stated in form of a conjecture.It hase to be proved in subsequent investigations.

MALKIN'S STABILITY RESULTS FOR CRITICAL NONLINEAR SYSTEMS.

In the proofs of the latter statements 4 to 6 the following results, due to Malkin are used.

Given a control system of the form

$$\dot{w} = A'.w + W(w) \quad , \quad w \in \mathbb{R}^n \quad , \quad A' \in \mathbb{R}^{n,n} \quad , \tag{1}$$

W analytic in w and at least of second order in the components of w.

For simplicity A' is assumed to have exactly one vanishing (critical) eigenvalue and n-1 eigenvalues with negative real part (this assumption will be called "assumption 1" later). Then we have

$$\det A' = o \quad . \tag{2a}$$

There exists a nonsingular transformation

$$w = T.z \quad , \quad z^T := (y^T, x) \quad , \quad x \in \mathbb{R}^1 \quad , \quad y \in \mathbb{R}^{n-1} \quad , \quad z \in \mathbb{R}^n \tag{2b}$$

such that (1) takes the form

$$\begin{pmatrix} \dot{y} \\ \dot{x} \end{pmatrix} = \begin{pmatrix} A_{11} & A_{12} \\ o & o \end{pmatrix} . \begin{pmatrix} y \\ x \end{pmatrix} + \begin{pmatrix} Y'(y,x) \\ X'(y,x) \end{pmatrix} \quad , \qquad \begin{matrix} (3a) \\ (3b) \end{matrix}$$

where

$$\begin{pmatrix} A_{11} & A_{12} \\ o & o \end{pmatrix} = T^{-1}.A'.T \in \mathbb{R}^{n,n} \quad , \quad A_{11} \in \mathbb{R}^{n-1,n-1} \quad , \quad A_{12} \in \mathbb{R}^{n-1,1} \quad ,$$

$$, \ \det A_{11} \neq o \quad \text{and} \quad \begin{pmatrix} Y'(y,x) \\ X'(y,x) \end{pmatrix} = T^{-1}.W(\ T.\begin{pmatrix} y \\ x \end{pmatrix}) \ .$$

Equation (1) is called <u>critical system</u> and equation (3b) <u>critical equation</u> corresponding to (1).

Then due to Malkin and Liapunov (W.Hahn, 1967 and R.Reissig et al. 1969) the following statements may be proved:

LEMMA (1):

Let n=1. Then equations (3a) and (3b) take the form

$$\dot{x} = X'(x) \quad \text{resp.} \quad \dot{x} = g.x^m + g_2.x^{m_2} + \ldots \quad , \quad 2 < m < m_2 < \ldots \quad . \tag{4}$$

a) The trivial solution of (4) is asymptotically stable iff

(i) $g < o$ and

(ii) $m = 2k+1$, $k = 1,2,\ldots$.

b) The trivial solution of (4) is unstable iff

(i) $g > o$ or if

(ii) $m = 2k$, $k = 1,2,\dots$.

c) The trivial solution of (4) is stable but not asymptotically stable iff

$X'(x) = o$ for all x resp. if $g = g_2 = \dots = o$.

LEMMA (2) :

Let $n > 1$. Then there exists a transformation

$$y = H(r) \quad , \quad r \in \mathbb{R}^{n-1}$$

such that (3a) and (3b) take the form

$$\begin{pmatrix} \dot{r} \\ \\ \dot{x} \end{pmatrix} = \begin{pmatrix} A_{11} & o \\ & \\ o & o \end{pmatrix} \cdot \begin{pmatrix} r \\ \\ x \end{pmatrix} + \begin{pmatrix} Y(r,x) \\ \\ X(r,x) \end{pmatrix} \tag{5}$$

where:

(i) The trivial solutions of (3) and (5) have the same stability behavior and

(ii) $X(o,x) = g.x^m + g_2.x^{m_2} + \dots , \; 2 \leqslant m < m_2$ resp. (6)

$$Y_j(o,x) = g_{1j}.x^{m_{1j}} + g_{2j}.x^{m_{2j}} + \dots \quad , \; m < m_{1j} < m_{2j} \dots$$

$$Y = \begin{pmatrix} Y_1 \\ \vdots \\ Y_{n-1} \end{pmatrix} \quad , \; j = 1,2,\dots,n-1 \; .$$

If $X(o,x) = o$ for all x the critical system (1) resp. (5) is called <u>singular.</u>

THEOREM (3):

Let (3a) and (3b) be of the special form (5a),(5b) and (6).

Then the statement of LEMMA(1) holds for this system.

AUXILIAR NOTATIONS,DEFINITIONS AND TECHNIQUES.

In order to construct a connexion between Malkin's stability results and special statements from bifurcation theory,the following definitions resp. notations turn out to be useful.

DEFINITION (1):

a) A system of the form

$$\dot{w} = A'.w + k^{\alpha}.B'.w + W(w) \quad ,k \in \mathbb{R}^1 \quad \text{parameter,} \quad \alpha = 1,2,3,\ldots \quad (7)$$

is called linear parametrically perturbed and related to system (1) (l.p.p.r (1)).

b) A system of the form

$$\dot{w} = A'.w + k^{\alpha}.B'.w^r + W(w) \quad , r = 2,3,\ldots, \quad (7')$$

$$w^r := \begin{pmatrix} w_1^r \\ \vdots \\ w_n^r \end{pmatrix} \quad , \quad w_j^r := \prod_{k=1}^{n} w_k^{\mu_{kj}} \, , \quad j = 1,2,\ldots,n; \, \mu_{kj} = 0,1,2,\ldots; \, \sum_{k=1}^{n} \mu_{kj} \geq 2$$

is called nonlinear parametrically perturbed and related to (1) (n.p.p.r. (1)).

c) A system of the form

$$\dot{w} = A'.w + \sum_{i,j=1,2,\ldots} B'_{ij}.k^i.w^j + W(w) \quad (7'')$$

is called multiterm parametrically perturbed and related to (1) (m.p.p.r. (1)).

In analogy to equation (1) the system (7) may be transformed by means of (2b) to "normal linear form" (8).

$$\begin{pmatrix} \dot{Y} \\ \dot{x} \end{pmatrix} = \begin{pmatrix} A_{11} & A_{12} \\ 0 & 0 \end{pmatrix} . \begin{pmatrix} Y \\ x \end{pmatrix} + k^{\alpha} . \begin{pmatrix} B_{11} & B_{12} \\ B_{21} & B_{22} \end{pmatrix} . \begin{pmatrix} Y \\ x \end{pmatrix} + \begin{pmatrix} Y'(Y,x) \\ X'(Y,x) \end{pmatrix} \begin{matrix} (8a) \\ (8b) \end{matrix}$$

where

$$\begin{pmatrix} B_{11} & B_{12} \\ B_{21} & B_{22} \end{pmatrix} := T^{-1}.B'.T.$$

Equations (7') and (7'') are transformed the same way to (8') and (8''). The stationary solutions $Y_{st}(k)$ and $x_{st}(k)$ of (8a) and (8b) are defined as solutions of (9a) and (9b).

$$\begin{pmatrix} 0 \\ 0 \end{pmatrix} = \begin{pmatrix} A_{11} & A_{12} \\ 0 & 0 \end{pmatrix} . \begin{pmatrix} Y_{st} \\ x_{st} \end{pmatrix} + k^{\alpha} . \begin{pmatrix} B_{11} & B_{12} \\ B_{21} & B_{22} \end{pmatrix} . \begin{pmatrix} Y_{st} \\ x_{st} \end{pmatrix} + \begin{pmatrix} Y'(Y_{st},x_{st}) \\ X'(Y_{st},x_{st}) \end{pmatrix} . \begin{matrix} (9a) \\ (9b) \end{matrix}$$

Because of the assumption (2a) equation (9a) can be solved uniquely with respect to y_{st} as a function of x_{st} and k , where the solution

$$y_{st} = \psi(x_{st},k) \qquad (1o)$$

may be represented by means of a convergent power series in x_{st} and k (the convergence of this series is guaranteed by the implicit function theorem).

Inserting (1o) into (9b) yields

$$o = k^{\alpha}.B_{21}.\psi(x_{st},k) + k^{\alpha}.B_{12}.x_{st} + X'(\psi(x_{st},k),x_{st}) \ . \qquad (11)$$

Equation (11) is called <u>branching equation</u> of (8a) and (8b) in bifurcation theory. It corresponds to the critical equation (3b) in stability theory.

In what follows we are interested in stationary solutions of the systems (7) resp. (8) which branch off the trivial solution in (x,k) = (o,o).

DEFINITION (2):

a) The point (x,k) = (o,o) is called a branch-point of the system (8a) and (8b) iff

there exist at least two different stationary branching solutions $_{\varkappa}x_{st}(k)$ of the system (8) for all $k \in U_h(k)$, $k \neq o$

where

$_{\varkappa}x_{st}(o) = o$, $_{\varkappa}x_{st}(k)$ are analytic in k , $\varkappa = 1,2, \ldots , n$

and

$U_h(k) = \{k : |k| < \varepsilon , \ k > o \text{ or } k < o \ , \varepsilon > o\}$,

(compare figure 1a, 1b and 1c).

b) A point (x,k) = (o,k') is called branchpoint of a <u>trivial bifurcation</u> iff at k=k' each admissable x satisfies the corresponding branching-equation (11) (it is always assumed that the trivial solution $x(k) \equiv o$ for all k is another solution of (11), compare figure 1d).

In what follows the branch-point is assumed to be the point (x,k)=(o,o).

There exist different approaches to investigate the number, the stability and the structure of the set of solutions $x_{st} = x_{st}(k)$ of equation (11)

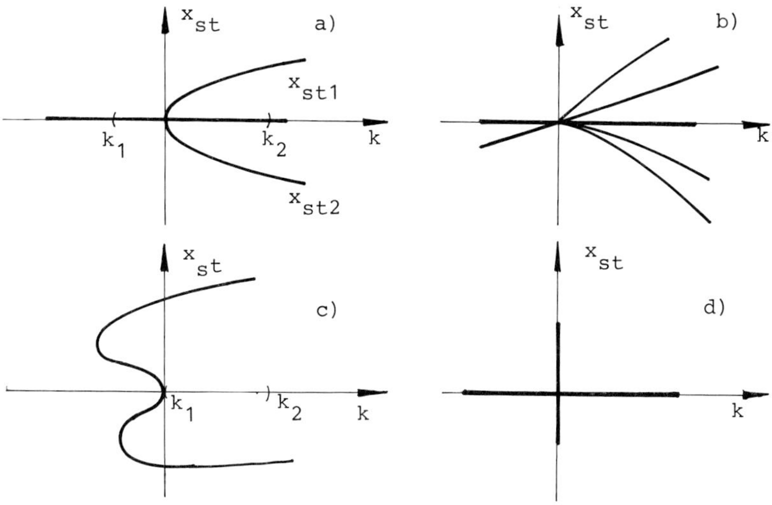

Figure 1: Branching-diagrams of stationary bifurcations $x=x_{st}(k)$

which bifurcate from a branch-point (branching-solutions).

In this paper results from a theory called "Newton's Diagram" (M.M.Vainberg and Trenogin,1974, H.Hahn et al.,1976) are used to construct the branching-solutions of (7) resp. (8).(The idea of this elegant and extremly useful technique dates back to the famous British physicist Sir Isaac Newton).

In case of a simple critical eigenvalue of the matrix A' (assumption 1) which implies that the branching equation (11) is a scalar equation, the different steps to find the branching-solutions of (11) may be collected in terms of the following rules:

Let $k^{\mu}.x_{st}^{\vartheta}$ be an arbitrary term of the branching-equation (11).

Rule 1: Draw a coordinate-system with ϑ as abscissa and μ as ordinate.

Rule 2: Insert the points corresponding to the exponents of each term of the branching-equation into this coordinate-system.

Rule 3: Draw the lowest convex polygon (convex from below) by connecting suitable points in the coordinate system (compare figure 2a).

Rule 4: The smallest distance (E) of the endpoint of this polygon from the ordinate of the coordinate-system (where only segments with negative slope are counted) is equal to the number of (real and complex) solutions of the branching-equation (11) which bifurcate from the point $(x,k) = (o,o)$.

Rule 5: The distance (S) of the startpoint of the polygon from the ordinate is equal to the multiplicity of the trivial solution $x_{st}(k) = o$ of (11).

Rule 6: The distance (L) of the end-point of the polygon from the abscissa is equal to the multiplicity of the trivial bifurcation (compare figure 2a).

Rule 7: In a neighbourhood of the branch-point $(x_{st},k) = (o,o)$ the branching solutions of (11) take the form

$$x_i(k) = \gamma_i . k^{\alpha_i} + \sum_{j=2} \gamma_{ij} . k^{\frac{p_{ij}}{q_i}} \text{ resp. } x_i(k) = \gamma_i . k^{\alpha_i} + o(|k^{\alpha_i}|), \quad (12a)$$

, where

$i=1,2,\ldots$; $\alpha_i := \tan \beta_i$, β_i as negative slope of the i-th segment in Newton's Polygon; $p_{ij}, q_i, \alpha_i \in \mathbb{Q}$, $\alpha_i < \frac{p_{ij}}{q_i}$.

Rule 8: Insert (12a) into (11) and compute the coefficients γ_i of (12a) (the corresponding equation is called supporting polynomial in bifurcation theory).

The higher order terms of (12a) are computed the same way. The application of these rules is demonstrated in the following example (compare figure 2b).

Example 1:

Given the following one dimensional branching equation (12b) :

$$o = k^5.x + k^4.x^3 + k^3.(x^2 + x^5 + x^6 + x^8) + k^2.(x^5 + x^7). \quad (12b)$$

According to the rules stated above, equation (12b) has the following branching solutions:

$$\begin{aligned} &x_o \equiv o \quad \text{for all } k \text{ (a simple solution)}, \\ &x_1 = \gamma_1 . k^{\alpha_1} + o(|k^{\alpha_1}|) \quad , \alpha_1 = \tan \beta_1 , \qquad (12c) \\ &x_{21} = \gamma_{21} . k^{\alpha_2} + o(|k^{\alpha_2}|) \quad , \alpha_2 = \tan \beta_2 , \\ &x_{22} = \gamma_{22} . k^{\alpha_2} + o(|k^{\alpha_2}|) \quad , \quad x_{23} = \gamma_{23} . k^{\alpha_2} + o(|k^{\alpha_2}|) \quad . \end{aligned}$$

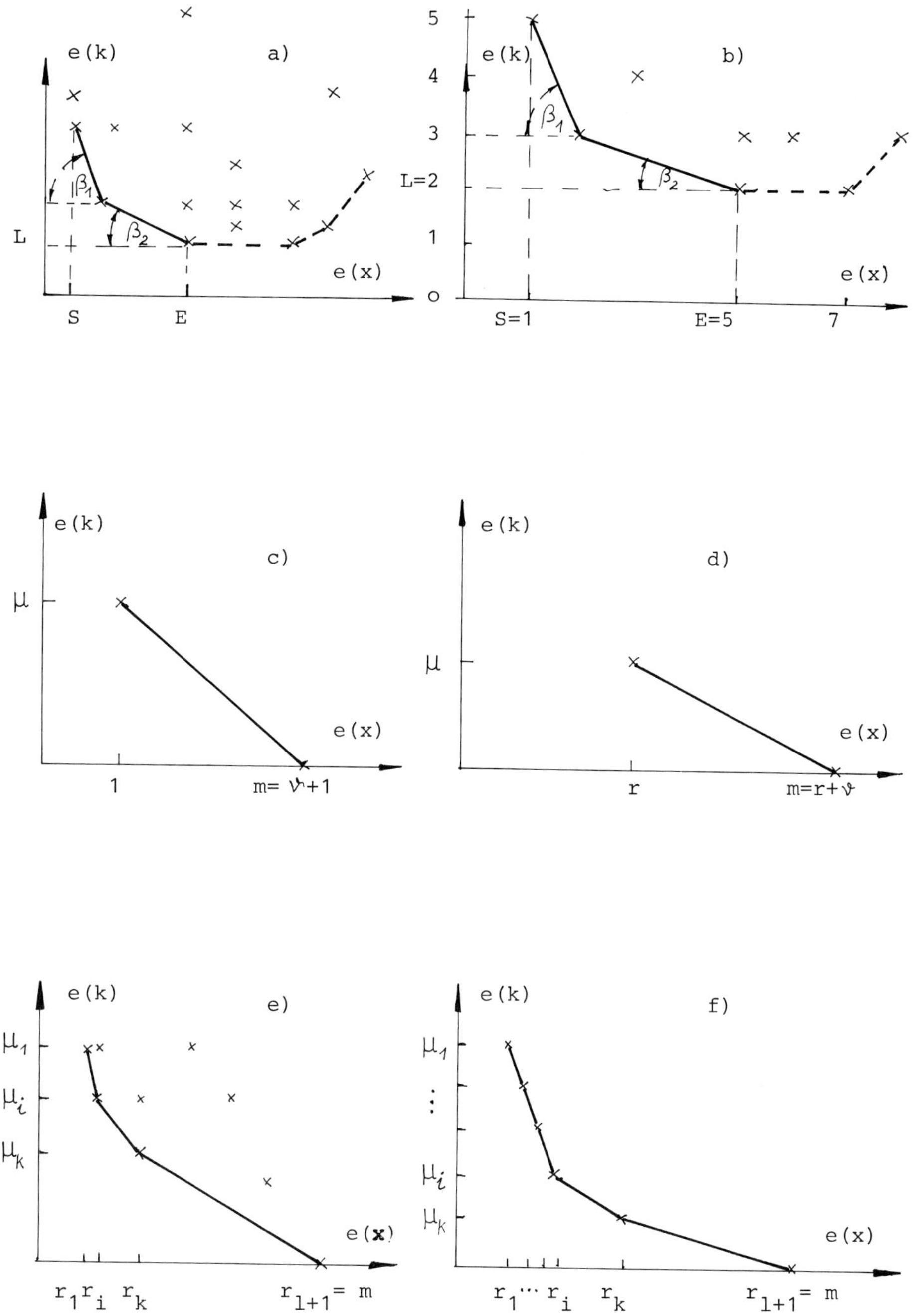

Newton-Diagrams corresponding to rule 1 to 7 (a), to equation (11a) (b), and to equations (13) (c), (13') (d) and (13'') (e) and (f), Figure 2 .

where

$$E = 5 \; , \; S = 1 \; , \; L = 2 \; , \; \alpha_1 = 2 \quad \text{and} \quad \alpha_2 = \frac{1}{3} \; .$$

The coefficients γ_1 and γ_{2i} (i=1,2,3) are solutions of the supporting polynomial

$$1 + \gamma_1 = o \quad \text{resp.}$$

$$1 + \gamma_2^3 = o \; . \tag{12d}$$

In what follows , the branching-equations (13),(13') and (13'') play a key role (as special cases of (11)).

Case 1 :

Given the branching equation

$$o = x.(b.k^{\mu} + g.x^{\vartheta}) \; ; \; b,g,k,x \in \mathbb{R}^1 \; ; \; \vartheta := m-1 \; ; \; \vartheta, \mu \in \mathbb{N} \; . \tag{13}$$

In the proof of the subsequent theorems the geometric structure of the real nontrivial solutions of (13) is needed for all combinations of the parameters b,g,μ and ϑ .
The extremely simple Newton-Diagram of (13) is sketched in figure 2c. Applying the corresponding rules, the branching solutions of (13) take the form

$$x_o(k) = o \quad \text{for all } k \; ,$$

$$x_1(k) = \gamma_i . k^{\frac{\mu}{\vartheta}} + o(|k^{\frac{\mu}{\vartheta}}|) \; , \; i=1,2, \; \ldots \; , \vartheta \; , \tag{14a}$$

where γ_i are solutions of the supporting polynomial

$$1 + \frac{g}{b}.\gamma^{\vartheta} = o \; . \tag{14b}$$

Equation (14b) has at most two different real roots. This implies that besides of the trivial solution at most two different branches bifurcate from the point (x,k) = (o,o). The qualitatively different graphs of the nontrivial real solutions (14a) of equation (13) are sketched in figures 3a and 3b for all combinations of the parameters b,g,μ and ϑ .

Case 2 :

Given the branching equation

$$o = x^r.(b.k^{\mu} + g.x^{\vartheta}) \; , \; r + \vartheta = m \; , \; r \in \mathbb{N} \; . \tag{13'}$$

$$b.k^{\mu} + g.x^{\nu} = o$$

$\underline{b.g < o}$

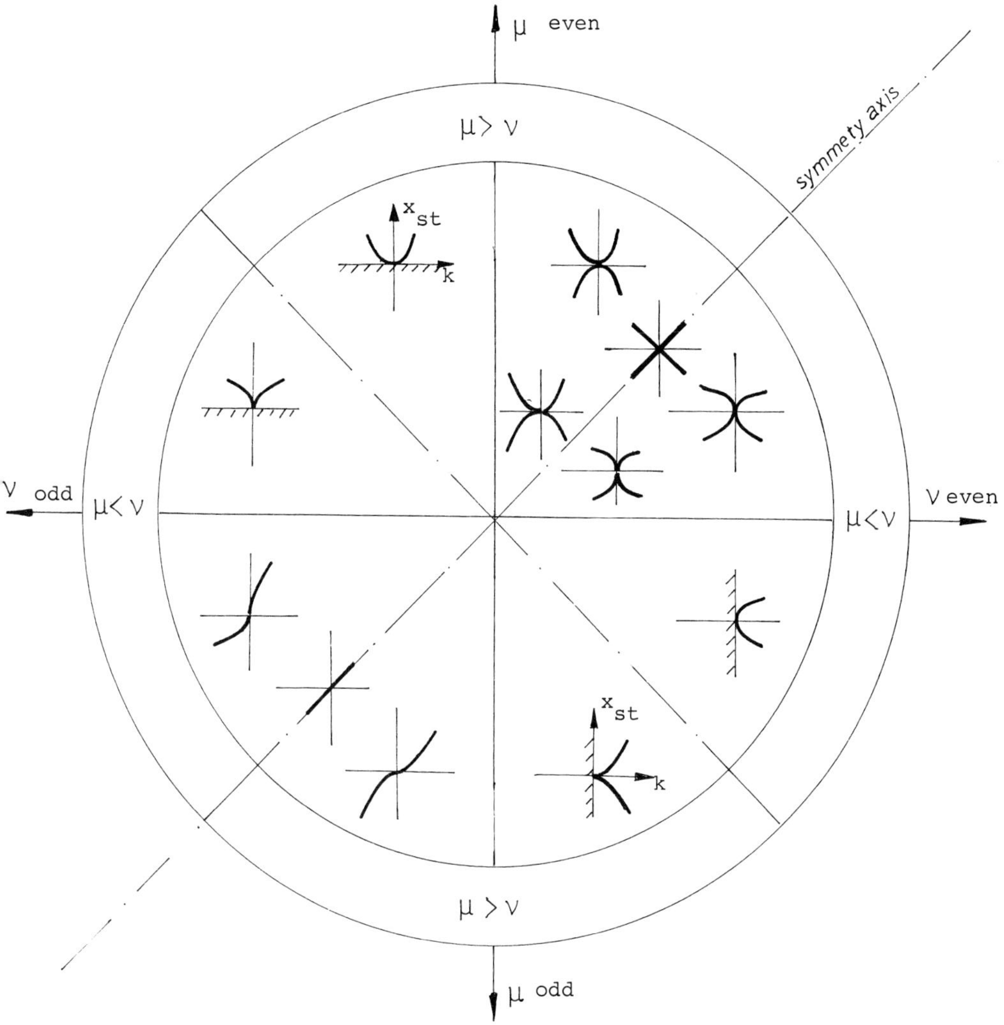

Figure 3a : Qualitative different graphs of the real solutions (14a).

$$b.k^{\mu} + g.x^{\nu} = o$$

$\underline{b.g > o}$

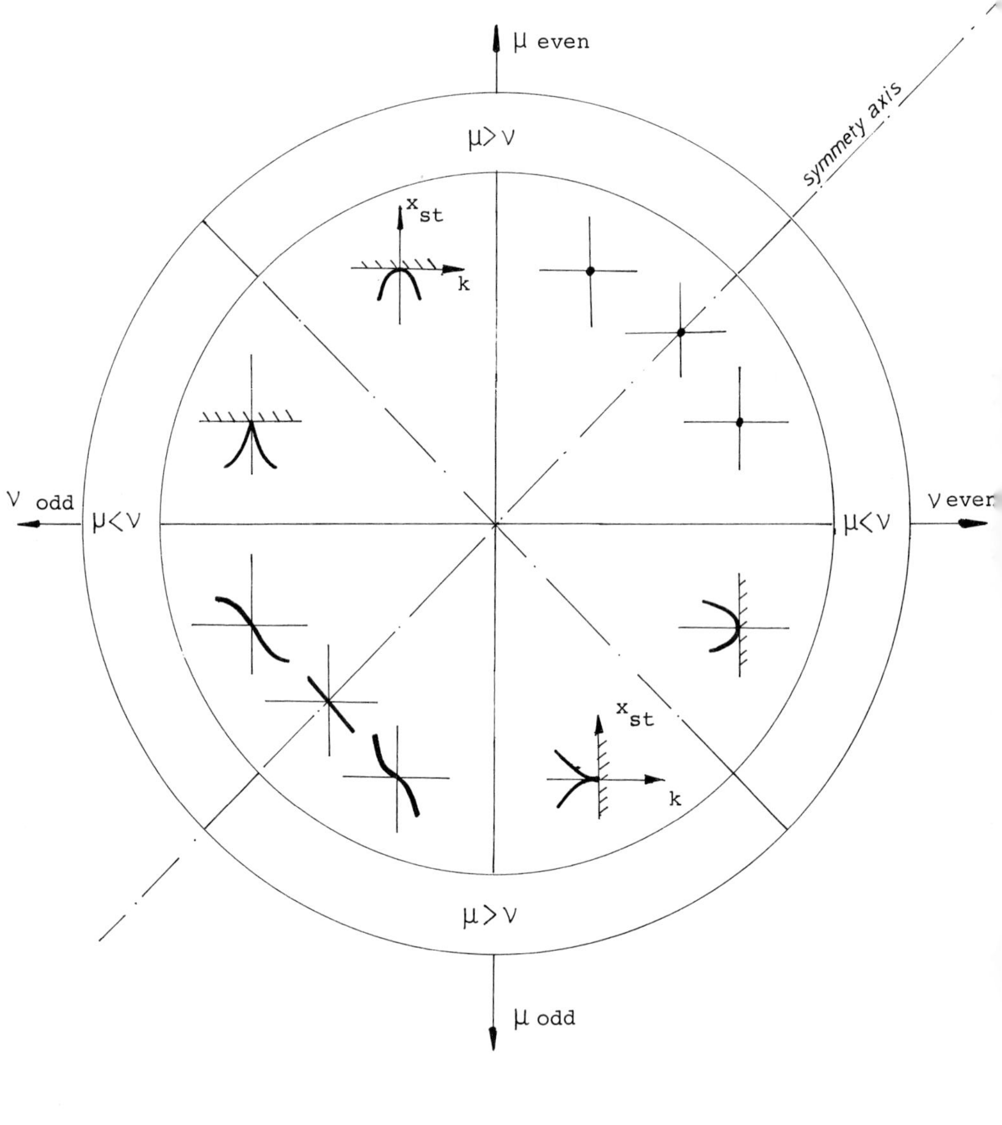

Figure 3b : Qualitative different graphs of the real solutions (14a)

Applying Newton's Diagram to equation (13') yields (compare figure 2d)

$$x_j(k) \equiv o \quad \text{for } j = 1,2,\ldots,r \quad ,$$
$$x_i(k) = \gamma_i \cdot k^{\frac{\mu}{\vartheta}} + o(|k^{\frac{\mu}{\vartheta}}|) \quad \text{for} \quad i = r+1, r+2,\ldots,m \quad , \tag{14a'}$$

where γ_i are solutions of the supporting polynomial

$$1 + \frac{g}{b} \cdot \gamma^{\vartheta} = o \quad . \tag{14b'}$$

The qualitative different graphs of the real solutions (14'a) are sketched in figures 3a and 3b (compare case 1).

Case 3:

Given the branching equation

$$o = \sum_{j=1}^{l+1} b_j \cdot k^{\mu_j} \cdot x^{r_j} \quad , \; b_{l+1} = g \; , \; r_{l+1} = m \; , \; \mu_{l+1} = o \; , \tag{13''}$$

$$r_{l+1} = m \qquad , \quad r_1 < \cdots < r_l < \quad m \quad \text{and} \quad \mu_j, r_j \in \mathbb{N} \; .$$

In connection with equation (13''), two different types of Newton-Diagrams may occur.

3.1 There exist at most l segments of the Newton-Polygon each of which contains exactly two points of the Newton-Diagram (as edge points), (compare figure 2e).

3.2 There exist exactly l-1 segments of the Newton-Polygon, and at least one of the segments contains more then two points from the Newton-Diagram (compare figure 2f).

In practical applications case 3.1 will occur more frequently. It may be analysed as follows:
Let the indices j in (13'') be relabelled in such a way, that the values j = 1,2,...,l+1 correspond to the points on the l segments of the Newton-Polygon, counted in natural order, i.e. corresponding to terms of (13'') with increasing exponents of the variable x. Then we have $r_{l+1} = m$, $b_{l+1} = g$ and $\mu_{l+1} = o$. As we are interested in branching solutions of (13'') in a neighbourhood of the point (x,k) = (o,o), the corresponding defining equations take the form (compare H. Hahn. 1976):

Segment 1 :

$$o = k^{\mu_2}.x^{r_1}(b_1.k^{\mu_1-\mu_2} + b_2.x^{r_2-r_1}) \quad , \tag{13.1}$$

$$x_{1i} = \gamma_{1i}.k^{\frac{\mu_1-\mu_2}{r_2-r_1}} + o(|q^1|) \, , \, q^j := k^{\frac{\mu_j-\mu_{j+1}}{r_{j+1}-r_j}} , \tag{14a.1}$$

$$o = 1 + \frac{b_2}{b_1}.\gamma_{1i}^{r_2-r_1} \quad . \tag{14b.1}$$

Segment 2 :

$$o = k^{\mu_3}.x^{r_2}(b_2.k^{\mu_2-\mu_3} + b_3.x^{r_3-r_2}) \quad , \tag{13.2}$$

$$x_{2i} = \gamma_{2i}.k^{\frac{\mu_2-\mu_3}{r_3-r_2}} + o(|q^2|) \quad , \tag{14a.2}$$

$$o = 1 + \frac{b_3}{b_2}.\gamma_{2i}^{r_3-r_2} \quad . \tag{14b.2}$$

$\vdots \qquad \vdots$

Segment l :

$$o = k^o.x^{r_l}(b_l.k^{\mu_l} + g.x^{m-r_l}) \quad , \tag{13.l}$$

$$x_{li} = \gamma_{li}.k^{\frac{\mu_l}{m-r_l}} + o(|q^l|) \quad , \tag{14a.l}$$

$$o = 1 + \frac{g}{b_l}.\gamma_{li}^{m-r_l} \quad . \tag{14b.l}$$

As already in the cases 1 and 2 , the supporting polynomials (14b.j), j=1,2,...,l, allow at most two real roots for each of the l segments.
Replacing μ resp. ϑ by means of $\bar{\mu}_i := \mu_i - \mu_{i+1}$ resp. by means of $\vartheta_i := r_{i+1} - r_i$ the geometric pattern of all nontrivial real solutions of (13.i) corresponding to the i-th segment of Newton's Polygon may be taken from figures 3a and 3b.
In order to get a complete picture of the real solutions corresponding to the whole Newton-Polygon, the different segments are to be

analyzed successively. This will be done in the proof of theorem (7).

The case 3.2 occurs much more seldom in applications. Here the number of terms of the supporting polynomial is equal to the number of points on the corresponding segment. As a consequence it is no longer as easy to compute the number of real branches resp. the bifurcation pattern of a segment with more than two points. This case will be omitted here.

In order to formulate the main theorems the following definition will be needed.

DEFINITION (3):

The trivial solution $x_{st}(k) \equiv o$ of (9a) and (9b) is called to be smoothly embedded on (the open interval) (k_1, k_2) iff

there exist two real stationary solutions ${}_1x_{st}(k)$ and ${}_2x_{st}(k)$ of the branching equation of (8) which bifurcate from the point $(x,k') = (o,o)$, $k' \in (k_1,k_2)$ such that

$$ {}_2x_{st}(k) < o < {}_1x_{st}(k) \quad \text{for all} \quad k \in (k_1,k_2) $$

and

$$ \lim_{k \to o} {}_ix_{st}(k) = o \quad , \quad i = 1,2 \quad \text{(compare figure 1a).} $$

Remark 1:

On assumption (1) -there exists exactly one simple critical eigenvalue of A'- the branching equation of the system (7) resp. (8) is a scalar equation, which may take the form of the equations (13),(13') or (13''). If the branching equation of (7) resp. of (8) takes the form (13), then the trivial solution of (8) is smoothly embedded in a dotted neighbourhood of k=o for μ even resp. in a dotted halfneighbourhood of k=o for μ odd, $b.\bar{g} < o$ and ϑ even (compare figure 3a).

Remark 2:

The concept of smooth embedding may be related to the definition (8a) (practical stability) in (H.Hahn, 1976 a). It should be noticed that branching diagrams corresponding to definition (6a) (practical jump stable) in(H.Hahn,a) are excluded from definition (3), (compare figure 1c).

Main results.

THEOREM (4):

The trivial solution $x(t) \equiv 0$ of the system (1) (resp. (3) or (5) is asymptotically stable (in m-th approximation) iff there exists a l.p.p.r.(1) system (8) *and a* corresponding branching equation (13) such that:

i) either the trivial solution of (7) resp. of (8) is (both) unstable and smoothly embedded in a dotted neighbourhood (for μ even) resp. in a dotted halfneighbourhood (for μ odd) of the point $(x,k) = (0,0)$; (compare figure 4a and 4b),

ii) or the trivial solution of (7) resp. of (8) is asymptotically stable in a dotted neighbourhood of $(x,k) = (0,0)$, where this latter point does not admit branching solutions (compare figure 4c).

(A dotted neighbourhood of a point P is defined as a neighbourhood of P with P removed).

Remark 3:

It is useful to notice that only l.p.p.r.(1) systems (7) or (8) may lead to a branching equation of (13), whereas only l.p.p.r. (1) systems (7) and n.p.p.r.(1) systems (7') may produce branching equations (13'). Branching equations (13'') may be produced by all types of perturbed systems (7), (7') and (7'').
If the trivial solution of (7) is asymptotically stable for $k > 0$ (resp. for $k < 0$) and unstable for $k > 0$ (resp. for $k < 0$) only one segment of the trivial solution (e.g. for $k < 0$) may be smoothly embedded by other solution branches. This will be different in situations of theorem (7) (m.p.p.r. (1) systems).

Proof :

In this proof, all the qualitative different combinations of the parameters b, g, μ und ϑ and the corresponding branching diagrams and stability pattern of the trivial solution of (11) resp. (13) as well as the corresponding stability behavior of (1) have to be investigated. As a consequence a lot of different cases occur. To get a clear perspective of all the different cases they will be listed in table I. The different columns resp. rows of table I have the following meaning:

In column 1 the trivial solution $x(k) \equiv 0$ of equation (7) is sketched (solid line for the asymptotical stable part and dotted

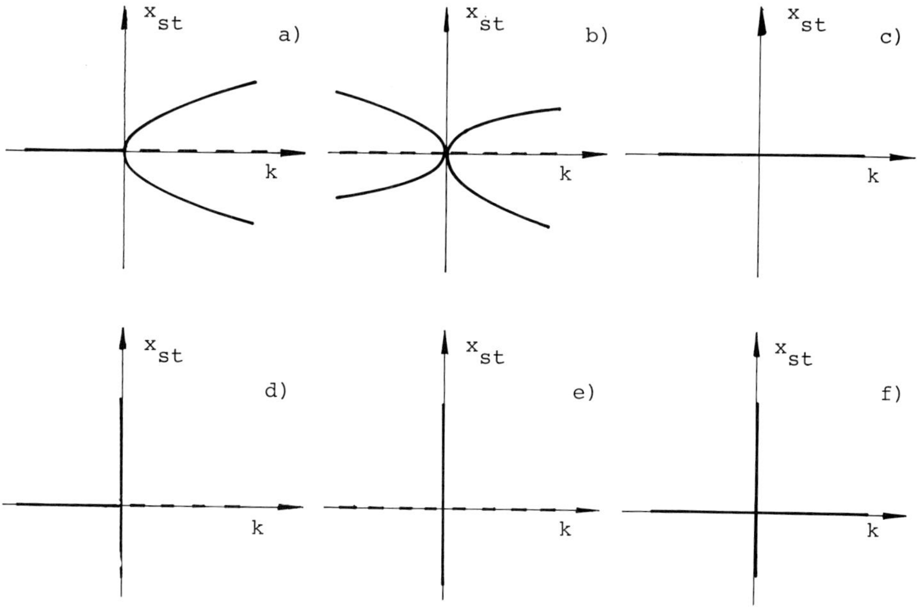

Figure 4 Branching-Diagrams corresponding to theorem (4), corollary (5) and to theorem (6) .

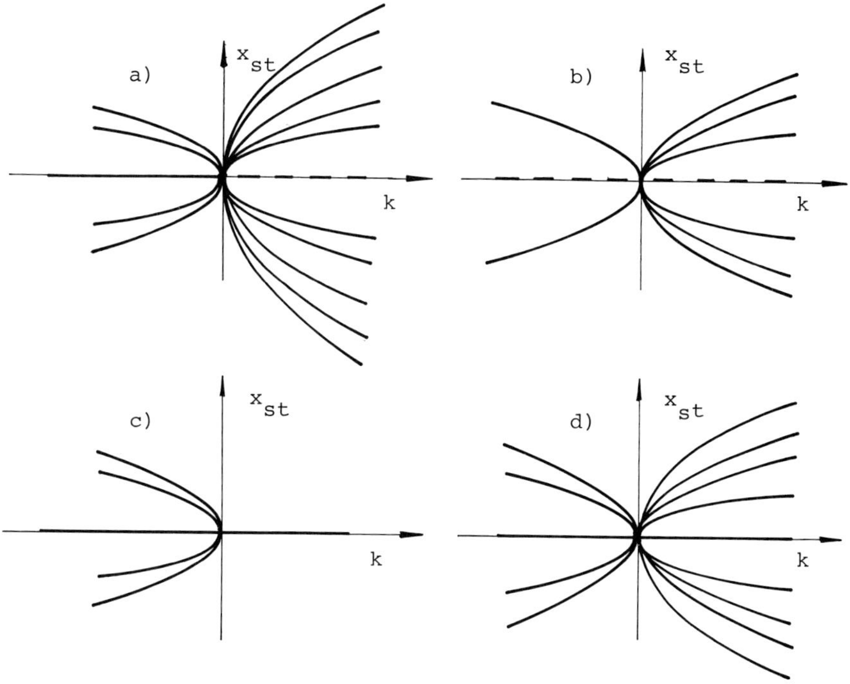

Figure 5 Branching-Diagrams corresponding to theorem (7) .

line for the unstable part of the solution).

In columns 2 and 4 the different branching diagrams, taken from figures 3a and 3b, are listed.

In columns 3 and 5 the different embedding diagrams corresponding to the trivial solution of (7) are drawn.

The rows 1 to 4 embrance the cases $b.g < o$.

The rows 5 to 8 are related to the case $b.g > o$.

$\longrightarrow$:

(i) Let the trivial solution of (1) be asymptotically stable. According to lemma (1),lemma (2) and theorem (3) this is the case iff $g < o$ and m is odd in (5) and (6) resp. in (13).Because of the relation $\vartheta = m-1$, ϑ is even. Then in case $b.g < o$ we have (because of $g < o$) $b > o$. For μ odd the trivial solution of (7) is unstable in first approximation and smoothly embedded in a dotted (positive) halfneighbourhood of k=o (for $k > o$);table I, column 3, row 1.
For μ even the trivial solution of (7) is unstable in first approximation and smoothly embedded in adotted neighbourhood of $k = o$ (for $k > o$ and for $k < o$);table I, column 3,row 2.
In the case $b.g > o$ we have $b < o$. For μ odd the trivial solution of (7) is unstable in first approximation and smoothly embedded in a dotted (negative) halfneighbourhood of k= o (for $k < o$).It is asymptotically stable for $k > o$; table I, column 3, row 5.
(ii) For $b.g > o$, $g < o$ and μ even the trivial solution of (7) is asymptotically stable in first approximation in a dotted neighbourhood of k=o (for $k > o$ and for $k < o$). This point does not allow bifurcations; table I, column 3, row 6.

All other cases either don't admit a smooth embedding of the unstable arcs of the trivial solution of (7) or-in case of an asymptotical stable trivial solution of (7) for all $k \neq o$- they don't admit any bifurcations.

$\longleftarrow$:

(i) and (ii) (Indirect proof). Assume the trivial solution of (1) is not asymtotically stable. Then it may be either unstable (a) or stable (b).
a) According to Malkin's theorems it is unstable iff either $g > o$ for arbitrary m or $g < o$ and m is even resp. ϑ (=m-1) is odd. Then any of the cases of table I, column 3, row 3, 4, 7 and 8 resp. column 5, row 1,2,3,4,5,6,7 and 8 may occur. Then either the unstable arcs of the

table I

l.p.p.r.			1 st t(7)	2 m odd∨ evenast t(1)	3	4 m even∨ oddi t(1)	5	
b.g < o	g<ob>o	μ odd						1
		μ even						2
	g>ob<o	μ odd		i t(1)		i t(1)		3
		μ even						4
b.g > o	g<ob<o	μ odd		ast t(1)		i t(1)		5
		μ even						6
	g>ob>o	μ odd		i t(1)		i t(1)		7
		μ even						8

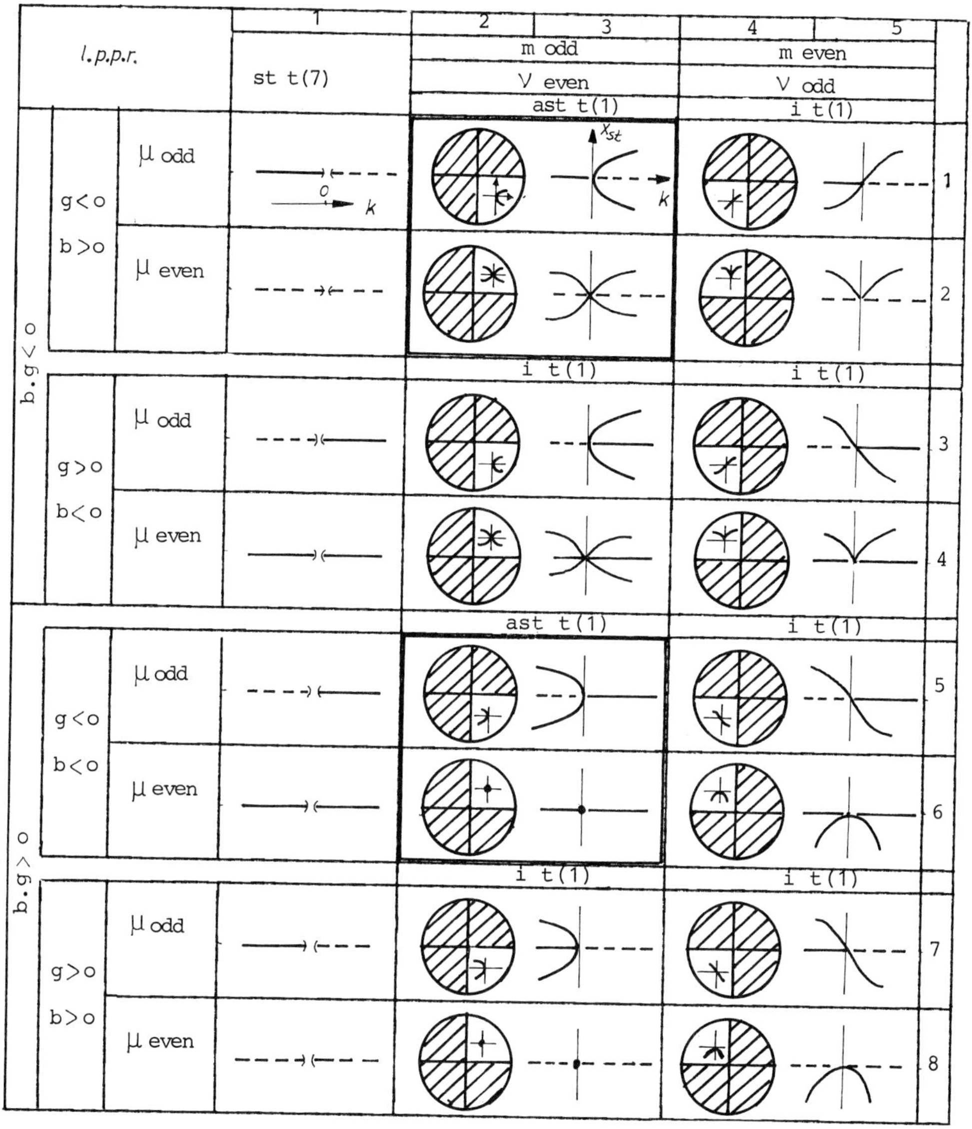

abbreviations:

st t(7) = the stability behaviour of the trivial solution of (7),

ast t(1)= the trivial solution of (1) is asymptotically stable,

i t(1) = the trivial solution of (1) is unstable,

─────── = an asymptotical stable part of the trivial solution of (7), (k ≠ o) ,

- - - - - = an unstable part of the trivial solution of (7), (k ≠ o).

trivial solution of (7) are not smoothly embedded (column 3, row 3,7 and 8 and column 5, row 1,2,3,5,7 and 8) or the trivial solution of (7) is asymptotically stable in a dotted neighbourhood of k=o (for $k > o$ and for $k < o$) and at least one branch bifurcates from this point; table I, column 3, row 4 and column 5, row 4 and 6.

b) Let the trivial solution of (1) be stable but not asymptotically stable. Then according to Malkin's theorems the singular case occurs which implies, that $X'(o,x) = o$ resp. $g = g_i = o$ for all i in (6). Then (13) is satisfied for k=o and for all x. According to definition (2) there exists only a trivial bifurcation which does not allow a smooth embedding of the trivial solution of (7).

COROLLARY (5):

The trivial solution of (1) (resp. (3) and (5)) is stable in m-th approximation but not asymptotically stable iff there exists a l.p.p.r. (1) system (11) or (13) such that the point $(x,k) = (o,o)$ only allows a trivial bifurcation (compare figure 4d,4e and 4f).

Proof :

$\rightarrow$:

Let the trivial solution of (1) be stable but not asymptotically stable in m-th approximation. Then according to the proof of theorem (4), paragraph b, the point $(x,k) = (o,o)$ only allows trivial bifurcations from the trivial solution of (7).

$\leftarrow$:

Let the trivial solution of (1) either be asymptotically stable or unstable in m-th approximation. Then according to table I there don't exist trivial bifurcations.

Remark 4:

It is an interesting fact, that by means of these theorems the stability behaviour of nonlinear critical systems may be decided by means of <u>pure algebraic</u> investigations (linear eigenvalue theory in combination with a discussion of the bifurcation pattern of the solutions of the algebraic branching equation). Moreover these criteria have a strong <u>geometric appeal.</u>

The results of theorem (4) and corollary (5) don't include any information with respect to the stability of the embedding (bifurcating) branches. The roots of this fact may be found in the principle of the exchange of stability of branching solutions (compare H.Hahn,1976b), which should also hold in a modified version in case of general situations associated to l.p.p.r.(1) systems as well as in case of n.p.p.r.(1) systems resp. of m.p.p.r.(1) systems, if additional conditons are fulfilled. It is easy to demonstrate, that this principle does not hold in its known form in case of n.p.p.r.(1) systems.

Remark 5:

It is useful to notice that (besides the technique of "Newtons Diagram") any suitable bifurcation theory may be used to decide the stability behaviour of nonlinear critical systems.
To be sure, the bifurcation theory of Liapunov and Schmidt (e.g.)-applied in (H.Hahn,1976b)- cannot be used to prove the statements of theorem (4) and corollary (5) as it only allows sufficient conditions for the existence of special bifurcations, whereas sufficient and necessary conditions are needed here.

In case of n.p.p.r.(1) systems the statements of theorem (4) only hold under additional assumptions. This will be shown in theorem (6).

THEOREM (6):

The trivial solution of (1) (resp. of (3)or (5)) is asymptotically stable in m-th approximation, iff there exists a l.p.p.r.(1) system (7) or a n.p.p.r.(1) system (7') together with a branching equation (13') and r odd, $o<r<m$, such that the statements (i) and (ii) of theorem (4) hold.

Proof:

The qualitative different stability pattern of the trivial solution of (7) resp. of (7') and (13') as well as the branching diagrams corresponding to all possible parameter values b,g,μ,ϑ,r and m ($b,g,\mu,\vartheta,r,m = o$ excluded) are collected in table II.

$\longrightarrow$:

(i) and (ii): Let the trivial solution of (1) be asymptotically stable. For r odd this will happen in case of m odd, $g < o$ and,because of $\vartheta = m - r$, ϑ even. The corresponding branching diagrams are collected in table II, column 2 and 3, row 1,2,9 and 1o. These branching diagrams are exactly the embedding diagrams of (i) and (ii) in theorem (4).

$\longleftarrow$:

(i) and (ii) : Assume the trivial solution of (1) is not asympto-

table II

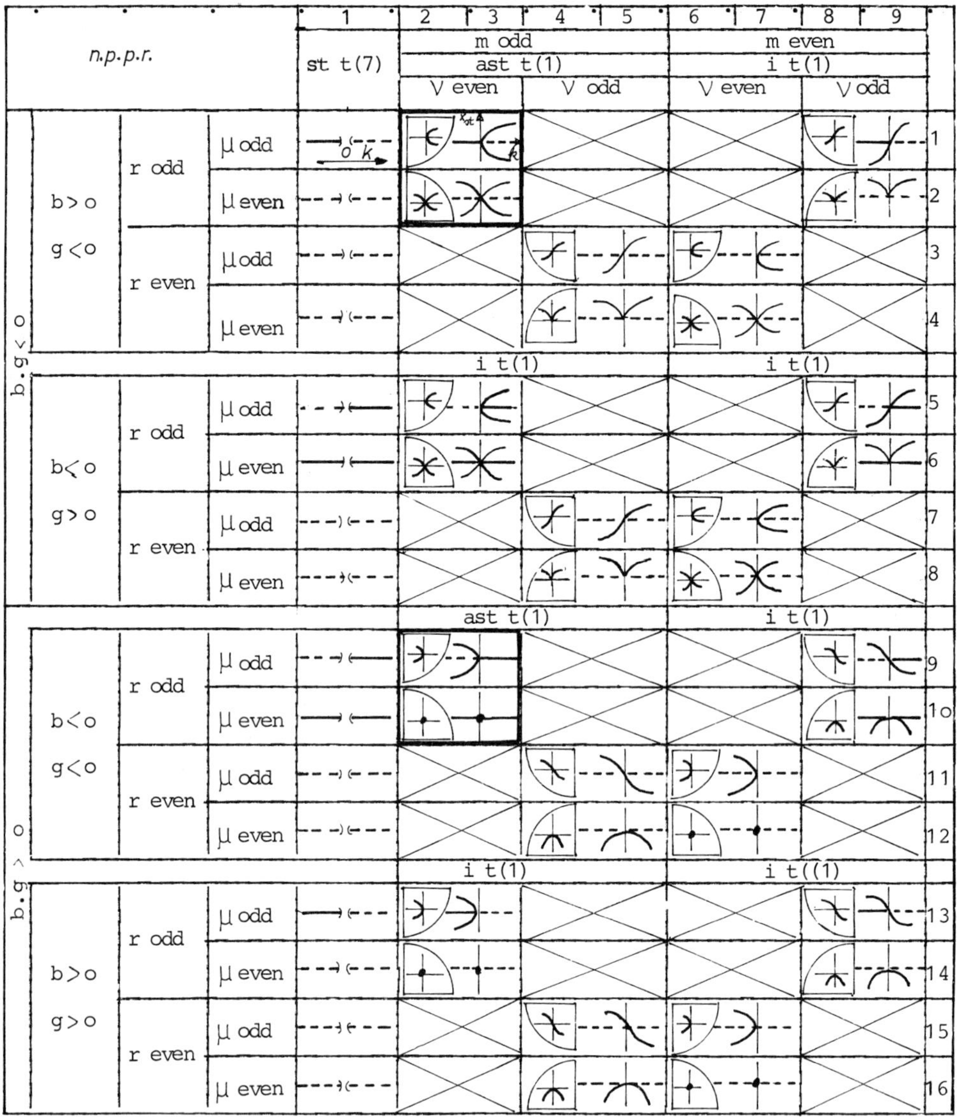

n.p.p.r.				1 st t(7)	2	3	4	5	6	7	8	9	
					m odd				m even				
					ast t(1)				i t(1)				
					∨ even		∨ odd		∨ even		∨ odd		
b.g < 0	b > o g < o	r odd	μ odd										1
			μ even										2
		r even	μ odd										3
			μ even										4
					i t(1)				i t(1)				
	b < o g > o	r odd	μ odd										5
			μ even										6
		r even	μ odd										7
			μ even										8
b.g > o					ast t(1)				i t(1)				
	b < o g < o	r odd	μ odd										9
			μ even										10
		r even	μ odd										11
			μ even										12
					i t(1)				i t((1)				
	b > o g > o	r odd	μ odd										13
			μ even										14
		r even	μ odd										15
			μ even										16

tically stable. Then two cases may occur: a) this solution is unstable or, b) this solution is stable but not attractive.

a) Let the trivial solution of (1) be unstable, $o < r$, r odd. Then two cases may occur:

a.1) g arbitrary and m even. Because of $\vartheta = m-r$ we have ϑ odd. The corresponding branching diagrams may be found in table II, column 8 and 9, row 1,2,5,6,9,1o,13 and 14.

a.2) $g > o$ and m odd. Then because of ϑ even, the corresponding branching diagrams may be found in table II, column 2 and 3 ,row 5,6,13 and 14.

In both cases embedding diagrams corresponding to (i) and (ii) in theorem (4) are excluded.

b) Let the trivial solution of (1) be stable but not attractive. Then according to Malkin's results $g = o$, which implies that there exists a trivial bifurcation of multiplicity μ. Again the conditions (i) and (ii) are hurt (this conclusion holds for r even and for r odd).

Remark 6:

It is easily shown, that for r even the statement of theorem (6) in general does not hold. For instance in case of column 5,row 3 of table II the trivial solution of (1) is asymptotically stable, but the unstable part of the trivial solution of (7') resp. of (13') is not smoothly embedded and in case of column 7, row 8, the trivial solution of (1) is unstable whereas the unstable part of the corresponding solution of (7') resp. (13') is smoothly embedded.

Remark 7:

The statement of corollary (5) is valid in case of n.p.p.r.(1)systems without any additional assumption (for arbitrary r).

For $r \geqslant m$ eqution (13') has no trivial solution $x(k) \equiv o$ for all k. This leads to a situation different from the one investigated here.

It should be noticed that for r even, the trivial solution of (7') is unstable for all (non vanishing) values of the remaining set of coefficients in case of an asymptotical stable trivial solution of (1). (compare table II, row 3,4,7,8,11,12,15, and 16).

In case of a m.p.p.r.(1) system (7'') resp. (8'') and a corresponding branching-equation (13'') the following statement may be proved:

THEOREM (7) :

The trivial solution of (1) (resp. of (3) and (5)) is asymptotically stable in m-th approximation iff there exists a m.p.p.r.(1) system (7'') together with a branching equation (13'') and a corresponding

Newton-Diagram of case 3. 1[4] , and $r_1, \dots, r_l$ odd, such that
a) the unstable segments of the trivial solution of (7'') resp. (8'') are smoothly embedded in a dotted neighbourhood (resp. halfneighbourhood) of $(x,k) = (o,o)$ by means of $2.(1+2.n)$ branches, $n = o,1, \dots$ (compare figure 5),

and

b) the asymptotically stable segments of the trivial solution of (7'') are smoothly embedded in a dotted neighbourhood (resp. in a dotted halfneighbourhood) of $(x,k) = (o,o)$ by means of $4.n$ branches, $n = o,1, \dots$ (compare figure 5);
(n may be different for different segments ($k > o$ or $k < o$) of the trivial solution of (7'')).

([4] Each segment of the Newton Polygon contains exactly two points of the Newton-Diagram corresponding to equation (13'')).

Remark 8:

In agreement with corollary (5) a trivial bifurcation of (13'') exists iff $g = o$,i.e. iff the trivial solution of (1) is stable but not unstable. In case of (13'') there may exist additional nontrivial (embedding) branching solutions besides of the trivial bifurcation.

Proof:

Theorem (7) will be proved step by step by encreasing the number of segments from 1 to l, l a finite number (because any Newton-Polygon contains at most a finite number of segments). Instead of trying to derive a general formula for the occurence of embedding branching diagrams (which might be rather cumbersome and not of much use in practice), Newton-Polygons of encreasing length will be analysed, until the generating rules for a construction of the corresponding branching pattern are obvious, and no new tables of branching diagrams are needed besides those which have already been used in the preceeding steps.
In principle the different stability pattern and branching-diagrams of each segment of the Newtondiagram may be computed in three steps:

Step 1 : The stability behaviour of the trivial solution of (1) depends on the coefficients g and m (the corresponding result is valid for each segment of a given Polygon).

Step 2 : The stability behaviour of the trivial solution of (7'') is determined by the constants b_1 , μ_1 and r_1 (this result again holds for all segments of the Polygon).

Step 3 : The branching-diagrams corresponding to the different segments depend on the following constants:

Segment 1 (the <u>first</u> segment in the Newtonpolygon) is determined by the constants $b_1, b_2, \bar{\mu}_1 := \mu_1 - \mu_2$ and $\vartheta_1 := r_2 - r_1$.

Segment i (an arbritary <u>intermediate</u> segment of the Newton-polygon) is determined by the constants $b_i, b_{i+1}, \bar{\mu}_i = \mu_i - \mu_{i+1}$ and $\vartheta_i = r_{i+1} - r_i$.

Segment l (the <u>last</u> segment of the Newtonpolygon) is determined by the constants $b_l, g, r_l, \bar{\mu}_l = \mu_l$ and $\vartheta_l = m - r_l$.

Taking into account the various branching-diagrams of figures 3a and 3b with μ, ϑ and b.g replaced by $\bar{\mu}_i, \vartheta_i$ and $b_i.b_{i+1}$ the stability pattern and branching-diagrams corresponding to the segment i of a Newton-polygon of length l are collected in the tables III for all qualitative different combinations of the parameters.

But not all of these pattern are allowed for each segment of the Newtonpolygon. The allowed pattern depend upon the length of the polygon and on the position of the segment within the polygon. The main restrictions of possible branching diagrams of a segment i are the relations

$$\bar{\mu}_i := \mu_i - \mu_{i+1} \quad , \quad \vartheta_i := r_{i+1} - r_i \quad , \quad i=1,2,\ldots,l+1. \tag{16}$$

Now the generating scheme for the different branching diagrams will be investigated by a discussion of Newtondiagrams of different length.

<u>Case l=1</u>:

The possible branching diagrams of the tables IIIa1 and IIIb1 are severely restricted by the relations

$$b_2 = g \quad , \quad \bar{\mu}_1 = \mu_1 \quad \text{and} \quad \vartheta_1 = m - r_1 \; . \tag{17.1}$$

Only the following pattern are allowed for r_1 odd :

table IIIa1 , column 2 and 5 , row 1 and 4,
column 6 and 9 , row 9 and 12 and

table IIIb1 , column 6 and 9 , row 1 and 4,
column 2 and 5 , row 9 and 12.

table IIIa1

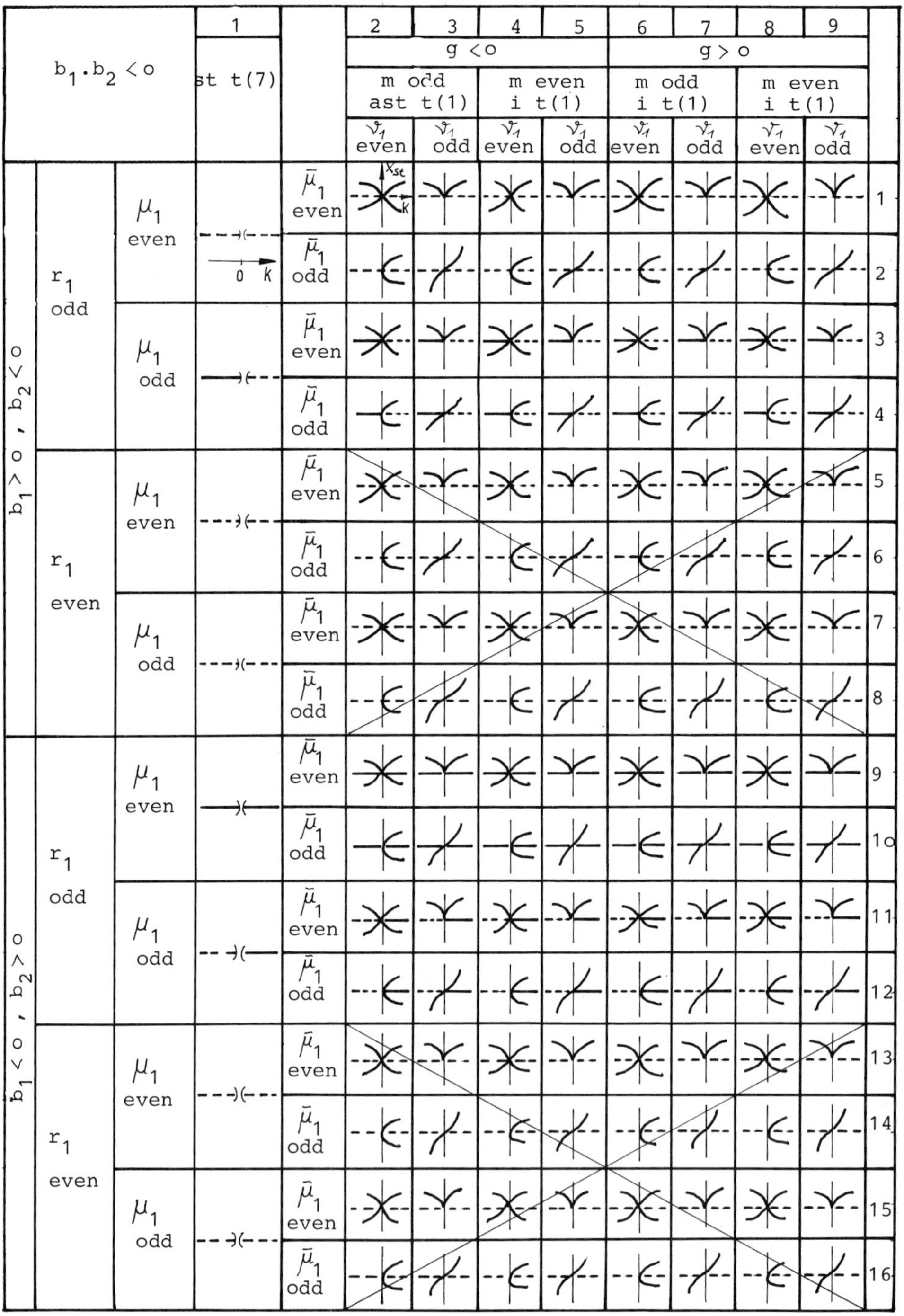

$b_1.b_2 < 0$			1		2	3	4	5	6	7	8	9	
			st t(7)		g < o				g > o				
					m odd ast t(1)		m even i t(1)		m odd i t(1)		m even i t(1)		
					ϑ_1 even	ϑ_1 odd	ϑ_1 even	ϑ_1 odd	ϑ_1 even	ϑ_1 odd	ϑ_1 even	ϑ_1 odd	
$b_1 > 0$, $b_2 < 0$	r_1 odd	μ_1 even		$\bar{\mu}_1$ even									1
			0 k	$\bar{\mu}_1$ odd									2
		μ_1 odd		$\bar{\mu}_1$ even									3
				$\bar{\mu}_1$ odd									4
	r_1 even	μ_1 even		$\bar{\mu}_1$ even									5
				$\bar{\mu}_1$ odd									6
		μ_1 odd		$\bar{\mu}_1$ even									7
				$\bar{\mu}_1$ odd									8
$b_1 < 0$, $b_2 > 0$	r_1 odd	μ_1 even		$\bar{\mu}_1$ even									9
				$\bar{\mu}_1$ odd									1o
		μ_1 odd		$\bar{\mu}_1$ even									11
				$\bar{\mu}_1$ odd									12
	r_1 even	μ_1 even		$\bar{\mu}_1$ even									13
				$\bar{\mu}_1$ odd									14
		μ_1 odd		$\bar{\mu}_1$ even									15
				$\bar{\mu}_1$ odd									16

table IIIb1

$b_1 \cdot b_2 > o$			1		2	3	4	5	6	7	8	9	
			st t(7)		$g < o$				$g > o$				
					m odd		m even		m odd		m even		
					ast t(1)		i t(1)		i t(1)		i t(1)		
					ϑ_1 even	ϑ_1 odd	ϑ_1 even	ϑ_1 odd	ϑ_1 even	ϑ_1 odd	ϑ_1 even	ϑ_1 odd	
$b_1 > o$, $b_2 > o$	r_1 odd	μ_1 even	0 K	$\bar{\mu}_1$ even	X_{st} k								1
				$\bar{\mu}_1$ odd									2
		μ_1 odd		$\bar{\mu}_1$ even									3
				$\bar{\mu}_1$ odd									4
	r_1 even	μ_1 even		$\bar{\mu}_1$ even									5
				$\bar{\mu}_1$ odd									6
		μ_1 odd		$\bar{\mu}_1$ even									7
				$\bar{\mu}_1$ odd									8
$b_1 < o$, $b_2 < o$	r_1 odd	μ_1 even		$\bar{\mu}_1$ even									9
				$\bar{\mu}_1$ odd									1o
		μ_1 odd		$\bar{\mu}_1$ even									11
				$\bar{\mu}_1$ odd									12
	r_1 even	μ_1 even		$\bar{\mu}_1$ even									13
				$\bar{\mu}_1$ odd									14
		μ_1 odd		$\bar{\mu}_1$ even									15
				$\bar{\mu}_1$ odd									16

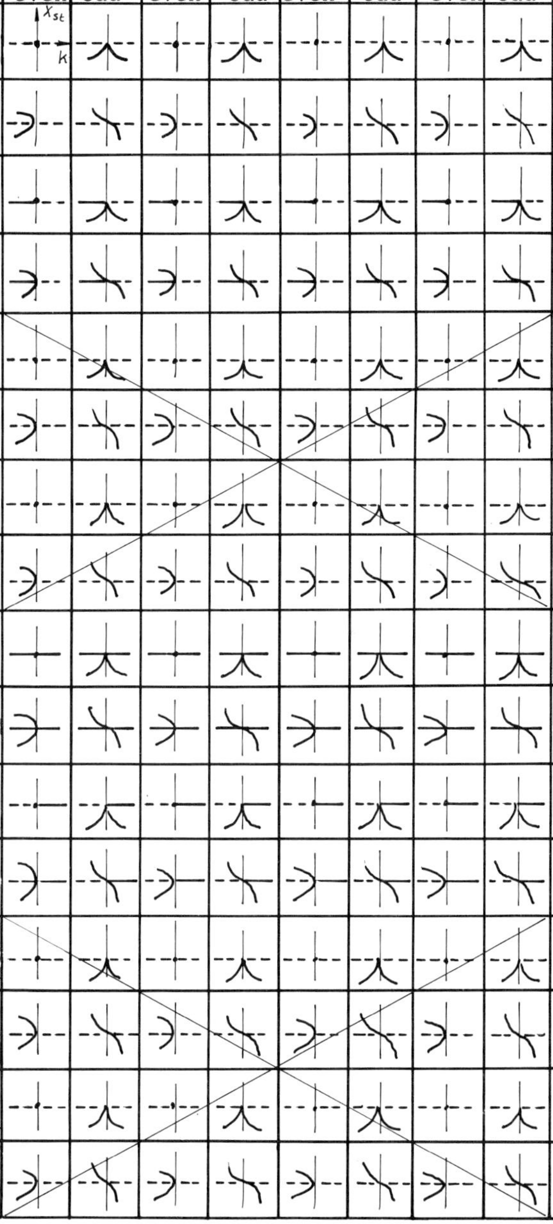

table IIIa2

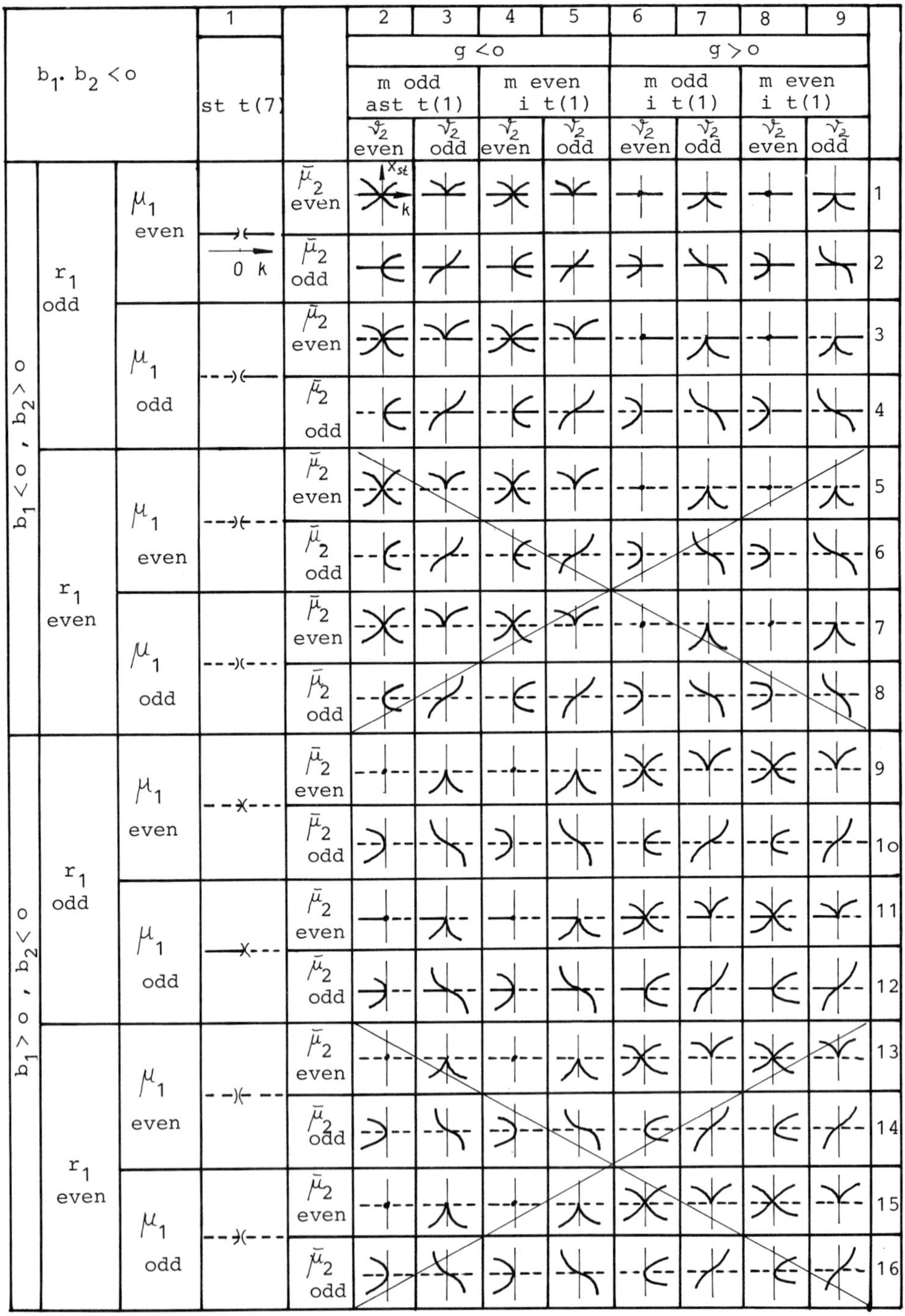

1
2
3
4
5
6
7
8
9
$b_1 \cdot b_2 < o$
st t(7)
g < o
g > o
m odd
ast t(1)
m even
i t(1)
m odd
i t(1)
m even
i t(1)
ϑ_2 even
ϑ_2 odd
ϑ_2 even
ϑ_2 odd
ϑ_2 even
ϑ_2 odd
ϑ_2 even
ϑ_2 odd
$b_1 < o$, $b_2 > o$
$b_1 > o$, $b_2 < o$
r_1 odd
r_1 even
r_1 odd
r_1 even
μ_1 even
μ_1 odd
μ_1 even
μ_1 odd
μ_1 even
μ_1 odd
μ_1 even
μ_1 odd
$\bar{\mu}_2$ even
$\bar{\mu}_2$ odd
x_{st}
k
0 k
1
2
3
4
5
6
7
8
9
1o
11
12
13
14
15
16

table IIIb2

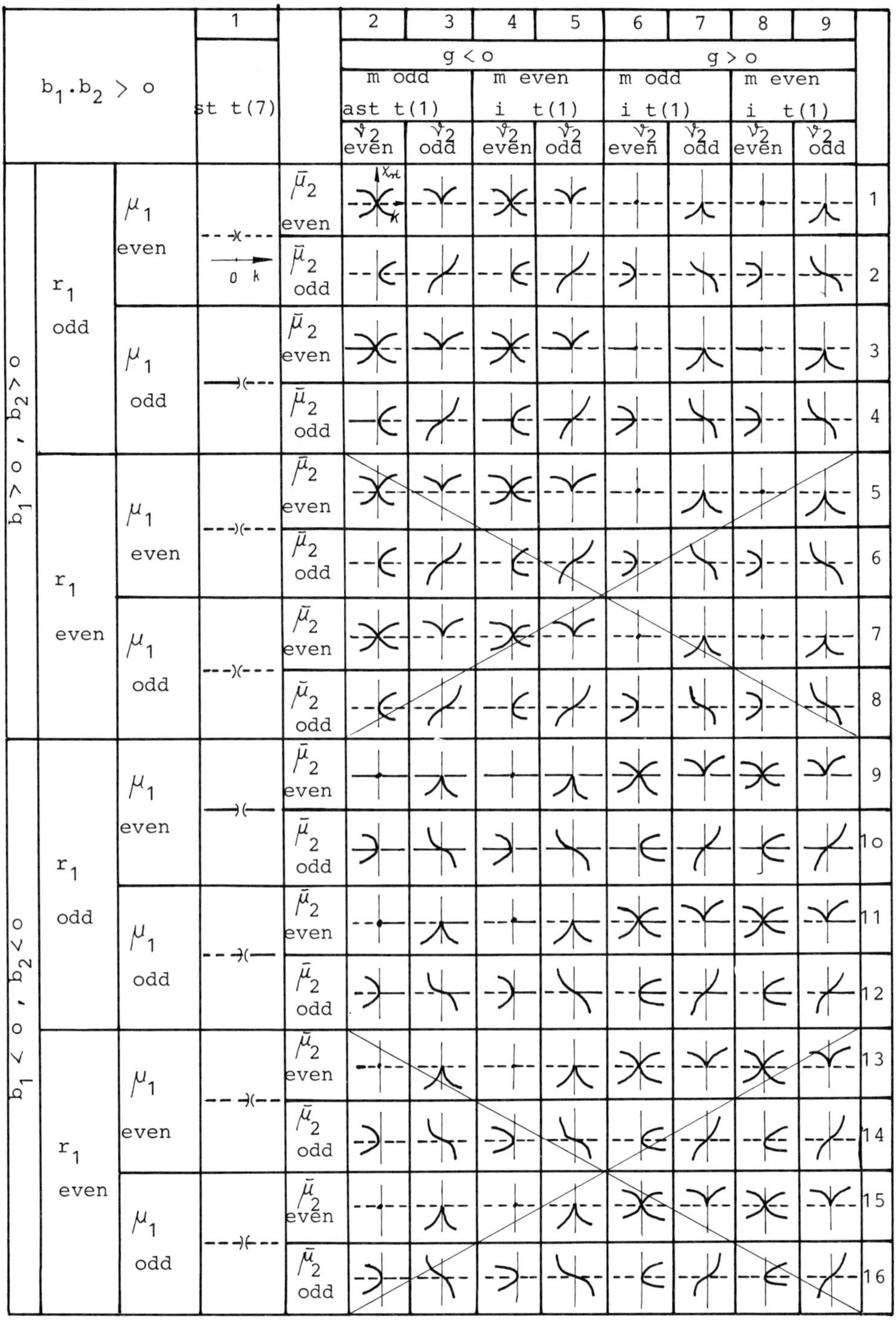

$b_1 \cdot b_2 > o$			1		2	3	4	5	6	7	8	9	
			st t(7)		g < o				g > o				
					m odd		m even		m odd		m even		
					ast t(1)		i t(1)		i t(1)		i t(1)		
					$\overset{v_2}{even}$	$\overset{v_2}{odd}$	$\overset{v_2}{even}$	$\overset{v_2}{odd}$	$\overset{v_2}{even}$	$\overset{v_2}{odd}$	$\overset{v_2}{even}$	$\overset{v_2}{odd}$	
$b_1 > o, b_2 > o$	r_1 odd	μ_1 even	$\bar\mu_2$ even										1
			$\bar\mu_2$ odd										2
		μ_1 odd	$\bar\mu_2$ even										3
			$\bar\mu_2$ odd										4
	r_1 even	μ_1 even	$\bar\mu_2$ even										5
			$\bar\mu_2$ odd										6
		μ_1 odd	$\bar\mu_2$ even										7
			$\bar\mu_2$ odd										8
$b_1 < o, b_2 < o$	r_1 odd	μ_1 even	$\bar\mu_2$ even										9
			$\bar\mu_2$ odd										10
		μ_1 odd	$\bar\mu_2$ even										11
			$\bar\mu_2$ odd										12
	r_1 even	μ_1 even	$\bar\mu_2$ even										13
			$\bar\mu_2$ odd										14
		μ_1 odd	$\bar\mu_2$ even										15
			$\bar\mu_2$ odd										16

These are just the allowed pattern of table II for r odd.
The embedding diagrams corresponding to asymptotical stable solutions of (1) are those of

table IIIa1, column 2, row 1 and 4 and
table IIIb1, column 2, row 9 and 12.

Case 1=2:

The possible pattern are listed in tables IIIa1, IIIb1, IIIa2 and IIIb2. The allowed pattern are restricted by the relations

$$b_3 = g \ , \ r_3 = m \quad \text{and} \quad (16) \text{ for } i = 1,2$$

where (17.2)

$$\bar{\mu}_2 = \mu_2 \quad \text{resp.} \quad \mu_3 = o \quad \text{and} \quad \nu_2 = m - r_2 \quad .$$

It is useful to start the analysis of the different segments of the Newtonpolygon beginning at the last segment (coefficients b_1 and g) and proceeding step by step to the first segment (coefficients b_1 and b_2).
Using the abbreviation "e" for even and "o" for odd, the relations (16) and (17.2) imply the allowed combinations of branching-diagrams listed in table IV for r_1 and r_2 odd and associated to an asymptotical stable trivial solution of (1).

table IV.1 , $b_1 . b_2 > o$, $b_1 > o$

$\mu_2 = \bar{\mu}_2$	μ_1	$\bar{\mu}_1$	branching diagram of segment 2	segment 1	both segments
e	e	e	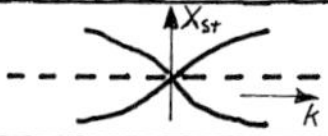		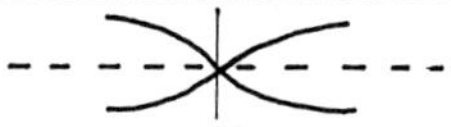
o	o	e			
e	o	o	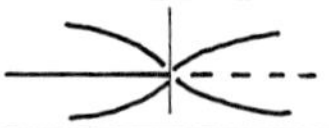	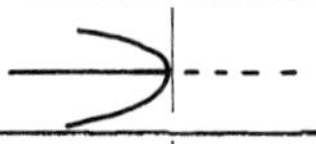	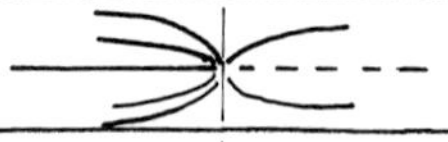
o	e	o	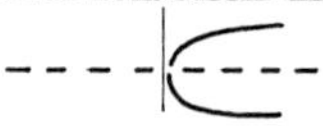	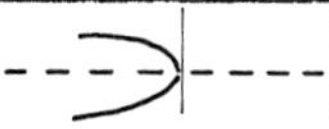	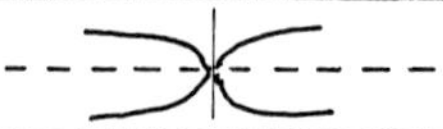

<u>table IV.2</u> , $b_1 \cdot b_2 > o$, $b_1 < o$

e	e	e
o	o	e
e	o	o
o	e	o

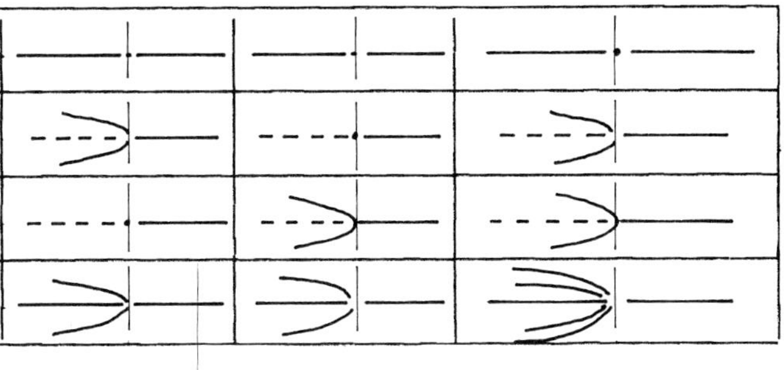

The branching diagrams of table IV.1 are taken from table IIIb.2 (segment 2) and table IIIb.1 (segment 1), column 2, row 1 to 4 . The branching diagrams of table IV.2 are taken from table IIIb.2 (segment 2) and table IIIb.1 (segment 1), column 2,row 9 to 12. The relations (16) and (17.2) guarantee the compatibility of the branching pattern corresponding to both segments.

<u>table IV.3</u> , $b_1 \cdot b_2 < o$, $b_1 > o$

$\mu_2 = \bar{\mu}_2$	μ_1	$\bar{\mu}_1$	branching diagram of segment 2	segment 1	both segments
e	e	e			
o	o	e			
e	o	o			
o	e	o			

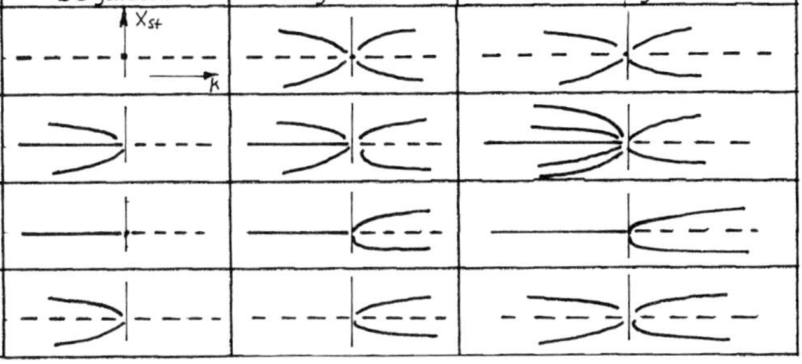

<u>table IV.4</u> , $b_1 \cdot b_2 < o$, $b_1 < o$

e	e	e
o	o	e
e	o	o
o	e	o

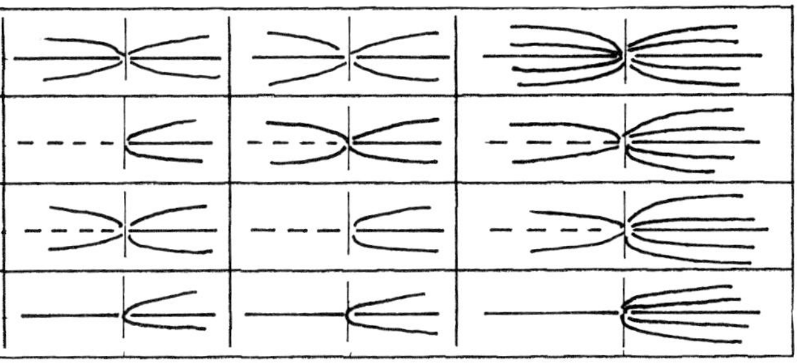

The branching diagrams of table IV.3 are taken from table IIIa.2, column 2, row 9 to 12 and from table IIIa.1, column 2, row 1 to 4.

The branching diagrams of table IV.4 are taken from table IIIa.2, column 2, row 1 to 4 and from table IIIa.1, column 2, row 9 to 12.

All branching diagrams of table IV.1 to IV.4 (both segments together) statisfy the embedding properties stated in theorem(7).

Applying the same procedure it is easily seen that these embedding properties are not always satisfied in case of r_1 and/or r_2 even. These cases have to be excluded from theorem (7). This is indicated by means of hatched lines in the various tables III.

Now the different branching diagrams corresponding to an unstable trivial solution of (1) will be discussed.
It is seen by inspection that the combined branching diagrams of the columns 4,5,8 and 9 of table IIIa.1, IIIa.2, IIIb.1 and IIIb.2 cannot produce a smooth embedding of the trivial solution of (8). The remaining columns 6 and 7 have to be investigated in analogy to the discussion of columns 2 and 3.
The allowed branching diagrams corresponding to an unstable trivial solution of (1) are listed in table V for m, r_1 and r_2 odd.

table V.1 , $b_1 . b_2 > 0$, $b_1 > 0$

$\mu_2 = \bar{\mu}_2$	μ_1	$\bar{\mu}_1$	branching diagram of segment 2	segment 1	both segments
e	e	e	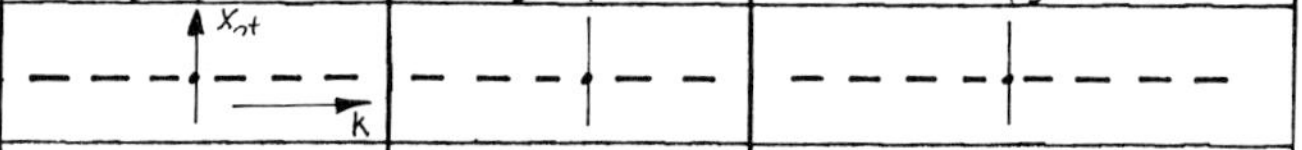		
o	o	e	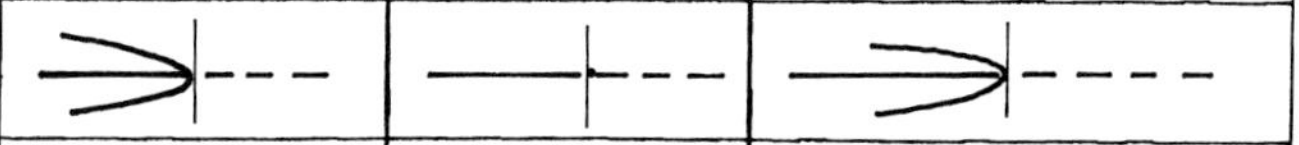		
e	o	o	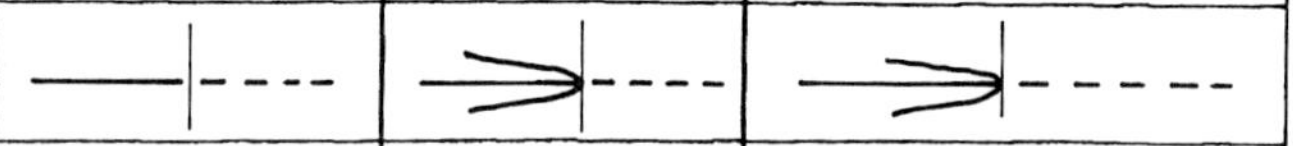		
o	e	o	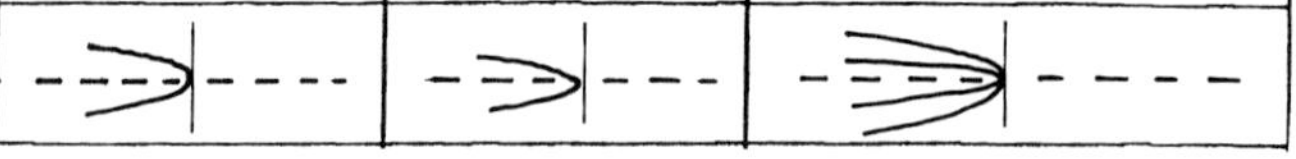		

<u>table V.2</u> , $b_1.b_2 > 0$, $b_1 < 0$

e	e	e
o	o	e
e	o	o
o	e	o

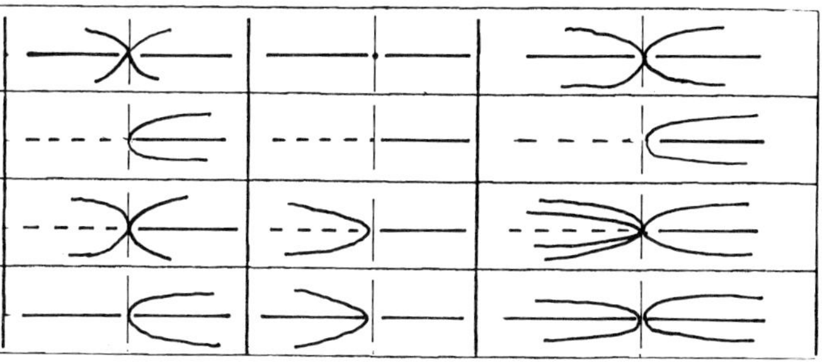

The branching diagrams of table V.1 (resp. of table V.2) are taken from table IIIb.2 and table IIIb.1, column 6, row 1 to 4 (resp. from table IIIb.2 and table IIIb.1, column 6, row 9 to 12).

<u>table V.3</u> , $b_1.b_2 < 0$, $b_1 > 0$

$\mu_2 = \bar{\mu}_2$	μ_1	$\bar{\mu}_1$
e	e	e
o	o	e
e	o	o
o	e	o

branching diagram of

segment 2 | segment 1 | both segments

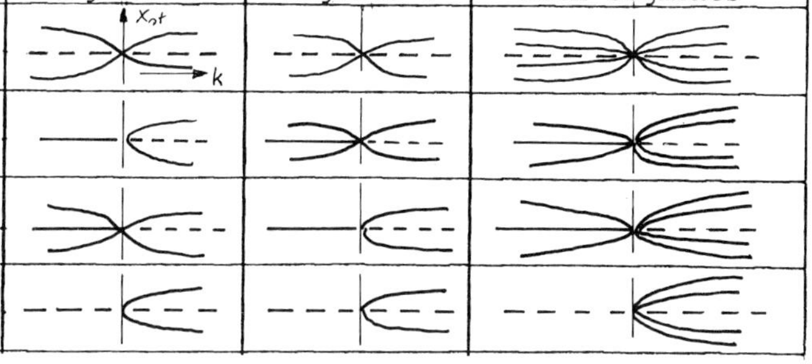

<u>table V.4</u> , $b_1.b_2 < 0$, $b_1 < 0$

e	e	e
o	o	e
e	o	o
o	e	o

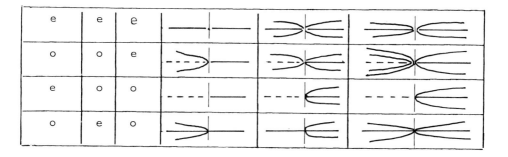

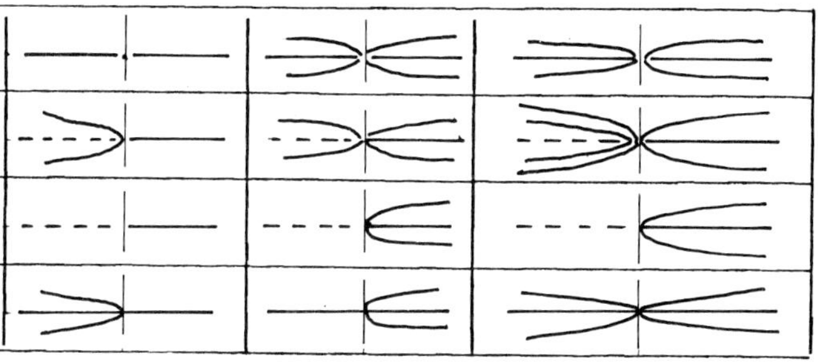

The branching diagrams of table V.3 (resp. of table V.4) are taken from table IIIa.2, column 6, row 9 to 12 and from table IIIa.1, column 6,row 1 to 4 (resp. from table IIIa.2 . column 6, row 1 to 4 and from table IIIa.1, column 6, row 9 to 12).

None of the branching diagrams of table V.1 to V.4 satisfies the embedding properties of theorem (7).

Applying the same procedure, it is easy to verify that these embedding properties of the various branching diagrams of table V.1 to V.4 are not always guaranteed for r_1 and /or r_2 even. This case must therefore be excluded from theorem (7).

Case l=3:

The possible stability pattern and branching diagrams are listed in table IIIa.1,IIIb.1,IIIa3.1,IIIb3.1,IIIa3.2,IIIb3.2,IIIa.2 und IIIb.2, where the index 2 in the latter two tables IIIa.2 and IIIb.2 is replaced by the index 3.

The allowed pattern are restricted by the relations

$$b_4 = g \ , \ r_4 = m \quad \text{and (16)} \quad \text{for } i = 1,2,3 \ , \quad \text{where}$$
$$\bar{\mu}_3 = \mu_3 \quad \text{resp.} \quad \mu_4 = o \quad \text{and} \quad \vartheta_3 = m - r_3 \ , \qquad (17.3)$$

and by the following combinations of coefficients b_j , j=1,2,3.

table VI

$b_1 \cdot b_3 > o$	$b_2 \cdot b_3 > o \rightarrow b_1 \cdot b_2 > o$	$b_1 > o, b_2 > o \rightarrow b_3 > o$
		$b_1 < o, b_2 < o \rightarrow b_3 < o$
	$b_2 \cdot b_3 < o \rightarrow b_1 \cdot b_2 < o$	$b_1 > o, b_2 < o \rightarrow b_3 > o$
		$b_1 < o, b_2 > o \rightarrow b_3 < o$
$b_1 \cdot b_3 < o$	$b_2 \cdot b_3 > o \rightarrow b_1 \cdot b_2 < o$	$b_1 > o, b_2 < o \rightarrow b_3 < o$
		$b_1 < o, b_2 > o \rightarrow b_3 > o$
	$b_2 \cdot b_3 < o \rightarrow b_1 \cdot b_2 > o$	$b_1 > o, b_2 > o \rightarrow b_3 < o$
		$b_1 < o, b_2 < o \rightarrow b_3 > o$

Using the abbreviation "e" for even and "o" for odd the relations (16) and (17.3) imply the allowed combinations of branching diagrams and stability pattern listed in table VII for r_1 and r_2 odd and associated to an asymptotical stable trivial solution of (1).

table IIIa3.1

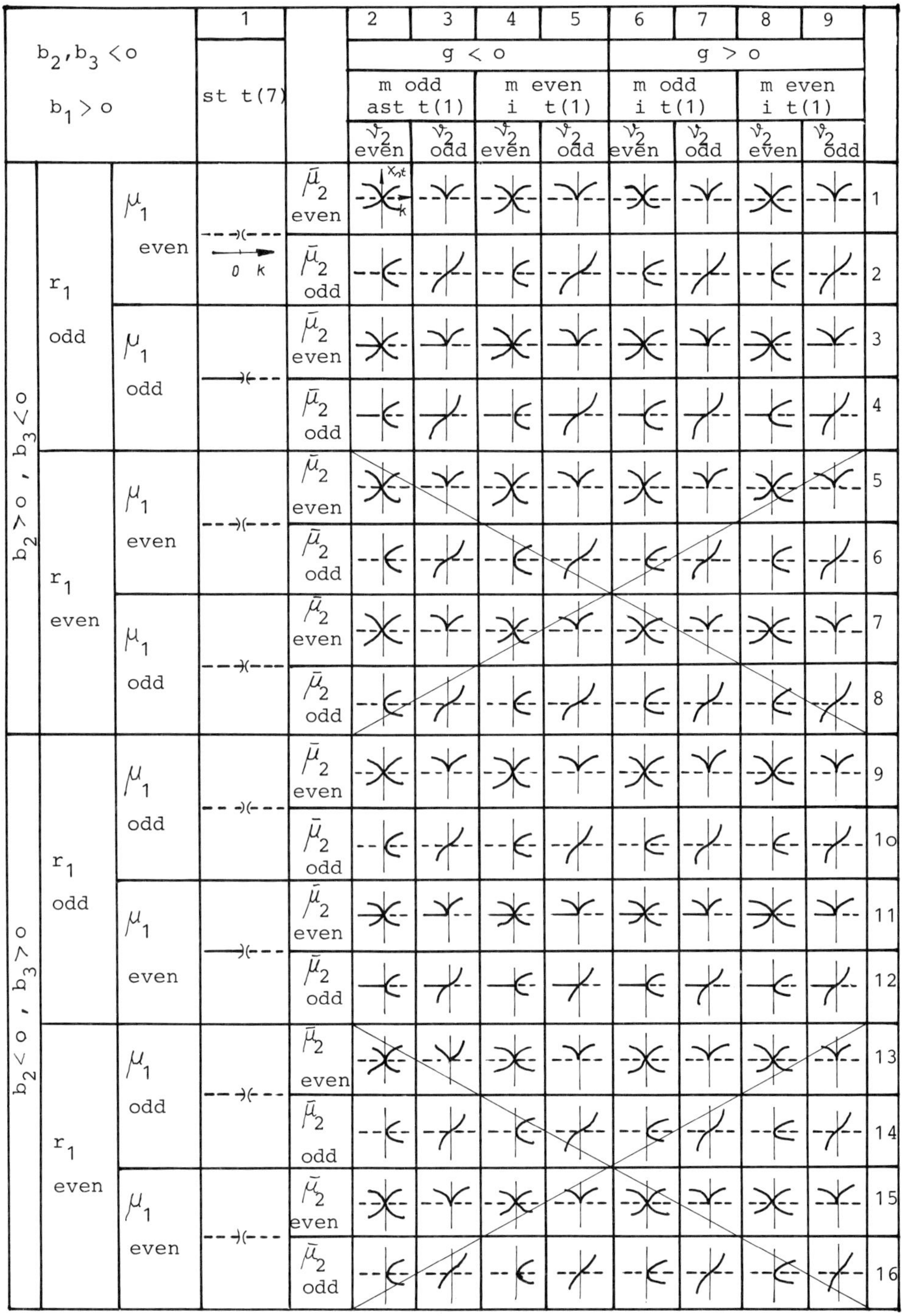
$b_2, b_3 < o$
$b_1 > o$
1
st t(7)
2
3
4
5
6
7
8
9
g < o
g > o
m odd
ast t(1)
m even
i t(1)
m odd
i t(1)
m even
i t(1)
ϑ_2 even
ϑ_2 odd
ϑ_2 even
ϑ_2 odd
ϑ_2 even
ϑ_2 odd
ϑ_2 even
ϑ_2 odd
$x_{n t}$
k
0 k
$b_2 > o$, $b_3 < o$
r_1 odd
μ_1 even
μ_1 odd
r_1 even
μ_1 even
μ_1 odd
$b_2 < o$, $b_3 > o$
r_1 odd
μ_1 odd
μ_1 even
r_1 even
μ_1 odd
μ_1 even
$\bar{\mu}_2$ even
$\bar{\mu}_2$ odd
1
2
3
4
5
6
7
8
9
1o
11
12
13
14
15
16

table IIIb3.1

$b_2 \cdot b_3 > o$ $b_1 > o$			1		2	3	4	5	6	7	8	9	
			st t(7)		$g < 0$				$g > 0$				
					m odd ast t(1)		m even i t(1)		m odd i t(1)		m even i t(1)		
					ν_2 even	ν_2 odd	ν_2 even	ν_2 odd	ν_2 even	ν_2 odd	ν_2 even	ν_2 odd	
$b_2 > 0$, $b_3 > o$	r_1 odd	μ_1 even		$\bar{\mu}_2$ even									1
				$\bar{\mu}_2$ odd									2
		μ_1 odd		$\bar{\mu}_2$ even									3
				$\bar{\mu}_2$ odd									4
	r_1 even	μ_1 even		$\bar{\mu}_2$ even									5
				$\bar{\mu}_2$ odd									6
		μ_1 odd		$\bar{\mu}_2$ even									7
				$\bar{\mu}_2$ odd									8
$b_2 < 0$, $b_3 < o$	r_1 odd	μ_1 even		$\bar{\mu}_2$ even									9
				$\bar{\mu}_2$ odd									1o
		μ_1 odd		$\bar{\mu}_2$ even									11
				$\bar{\mu}_2$ odd									12
	r_1 even	μ_1 even		$\bar{\mu}_2$ even									13
				$\bar{\mu}_2$ odd									14
		μ_1 odd		$\bar{\mu}_2$ even									15
				$\bar{\mu}_2$ odd									16

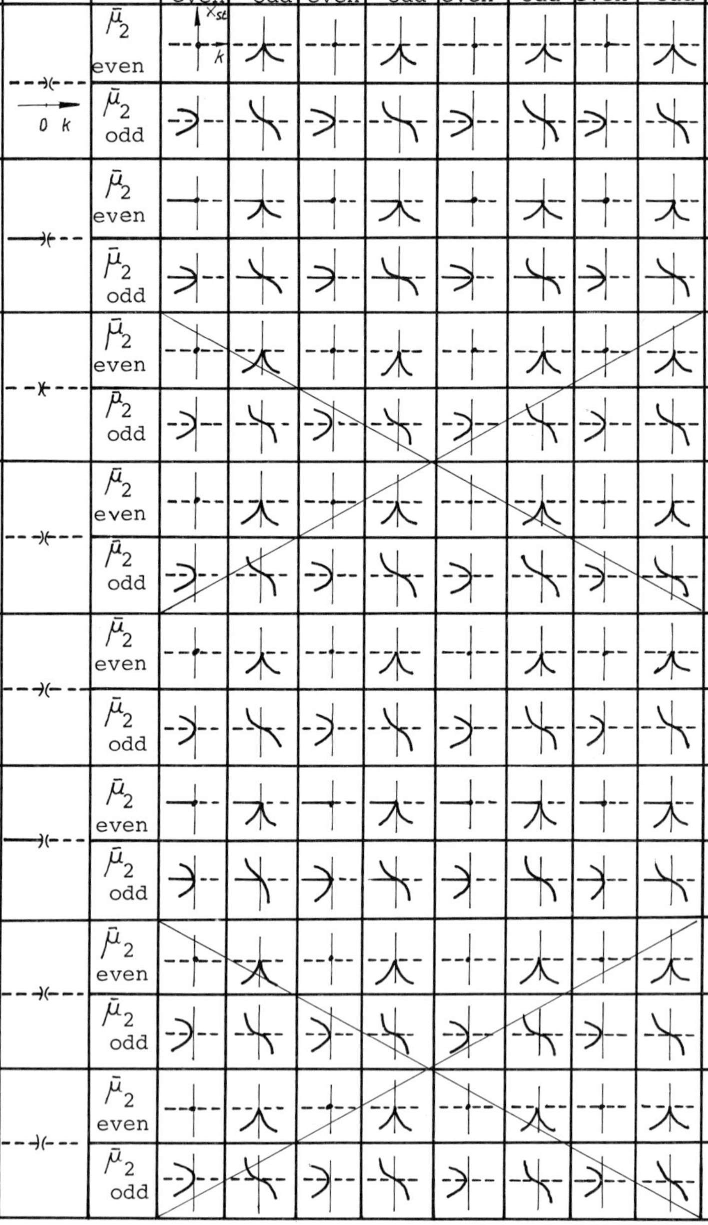

table IIIa3.2

$b_2 . b_3 < o$, $b_1 < o$			1		2	3	4	5	6	7	8	9	
			st t(7)		$g < 0$				$g > 0$				
					m odd, ast t(1)		m even, i t(1)		m odd, i t(1)		m even, i t(1)		
					ν_2 even	ν_2 odd	ν_2 even	ν_2 odd	ν_2 even	ν_2 odd	ν_2 even	ν_2 odd	
$b_2 > 0$, $b_3 < 0$	r_1 odd	μ_1 even		$\bar{\mu}_2$ even									1
				$\bar{\mu}_2$ odd									2
		μ_1 odd		$\bar{\mu}_2$ even									3
				$\bar{\mu}_2$ odd									4
	r_1 even	μ_1 even		$\bar{\mu}_2$ even									5
				$\bar{\mu}_2$ odd									6
		μ_1 odd		$\bar{\mu}_2$ even									7
				$\bar{\mu}_2$ odd									8
$b_2 < 0$, $b_3 > 0$	r_1 odd	μ_1 even		$\bar{\mu}_2$ even									9
				$\bar{\mu}_2$ odd									1o
		μ_1 odd		$\bar{\mu}_2$ even									11
				$\bar{\mu}_2$ odd									12
	r_1 even	μ_1 even		$\bar{\mu}_2$ even									13
				$\bar{\mu}_2$ odd									14
		μ_1 odd		$\bar{\mu}_2$ even									15
				$\bar{\mu}_2$ odd									16

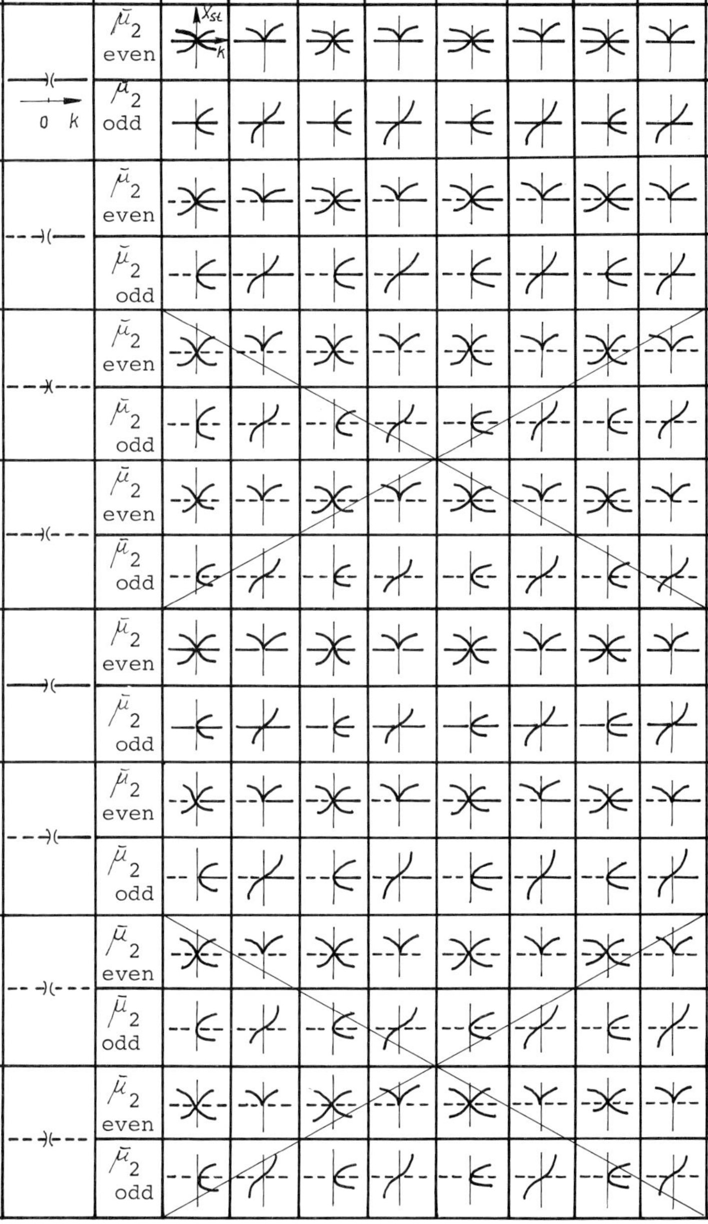

tableIIIb3.2

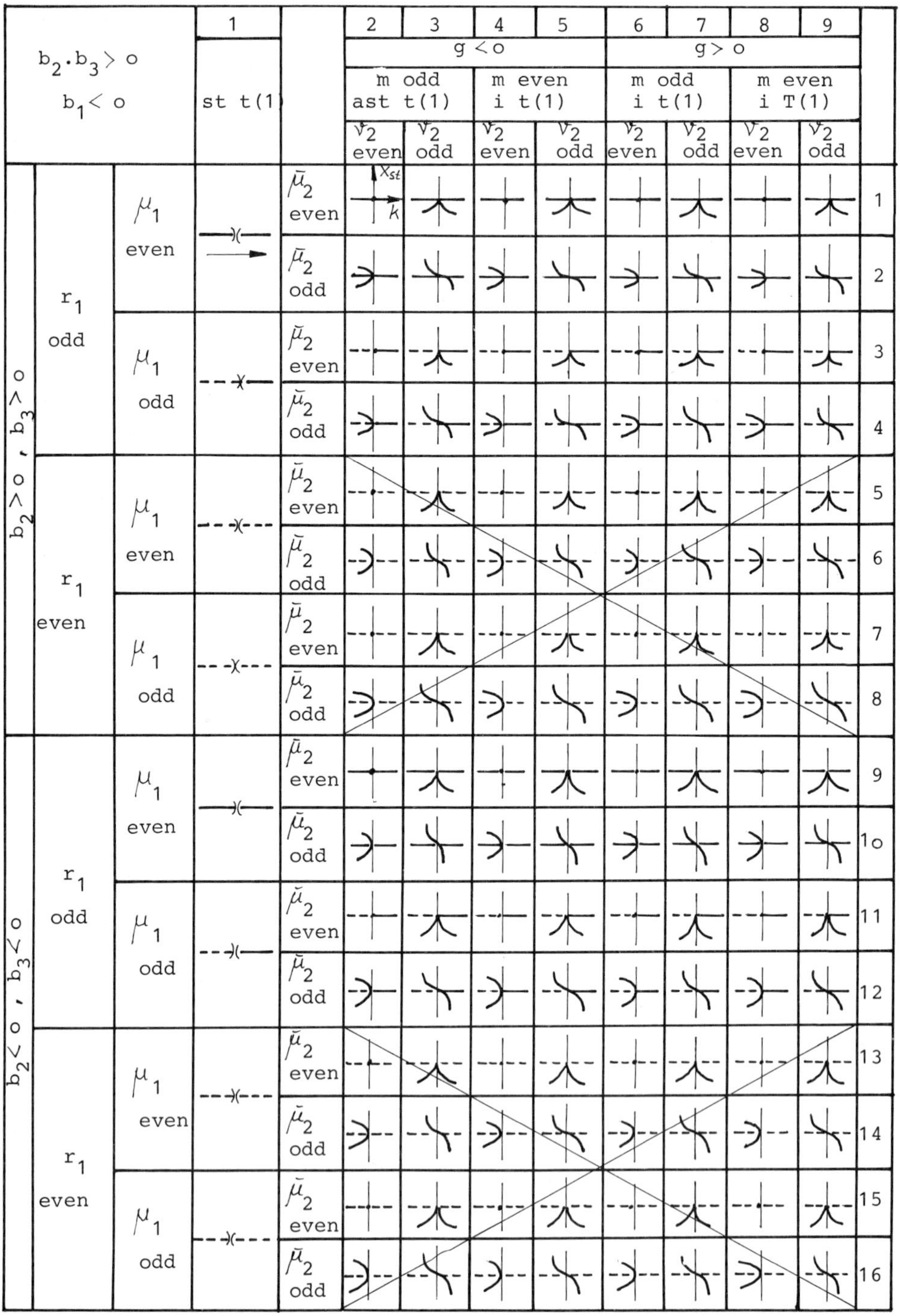
$b_2 \cdot b_3 > o$
$b_1 < o$
1
st t(1)
2
3
4
5
6
7
8
9
$g < o$
$g > o$
m odd
ast t(1)
m even
i t(1)
m odd
i t(1)
m even
i T(1)
ν_2 even
ν_2 odd
x_{st}
k
$b_2 > o$, $b_3 > o$
$b_2 < o$, $b_3 < o$
r_1 odd
r_1 even
μ_1 even
μ_1 odd
$\bar{\mu}_2$ even
$\bar{\mu}_2$ odd
1
2
3
4
5
6
7
8
9
1o
11
12
13
14
15
16

table VII.1 , $b_1.b_3 > o$, $b_2.b_3 > o$, $b_1.b_2 > o$

	μ_2	$\bar{\mu}_3$	μ_1	$\bar{\mu}_2$	$\bar{\mu}_1$	branching diagram of segment 3	segment 2	segment 1	all segments
$b_1 > o$, $b_2 > o$	e	e	e	e	e				
		o	o	o	o				
	o	e	e	o	o				
		o	o	e	e				
$b_1 < o$, $b_2 < o$	e	e	e	e	e				
		o	o	o	o				
	o	e	e	o	o				
		o	o	e	e				

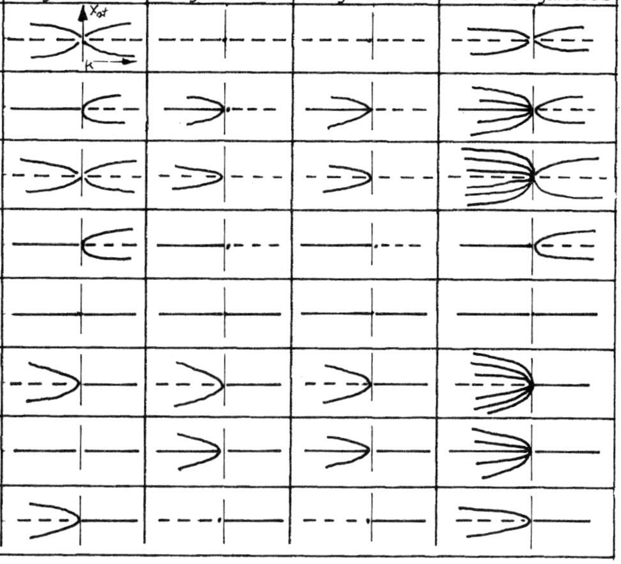

table VII.2 , $b_1.b_3 > o$, $b_2.b_3 < o$, $b_1.b_2 < o$

$b_1 > o$, $b_2 < o$	e	e	e	e	e				
		o	o	o	o				
	o	e	e	o	o				
		o	o	e	e				
$b_1 < o$, $b_2 > o$	e	e	e	e	e				
		o	o	o	o				
	o	e	e	o	o				
		o	o	e	e				

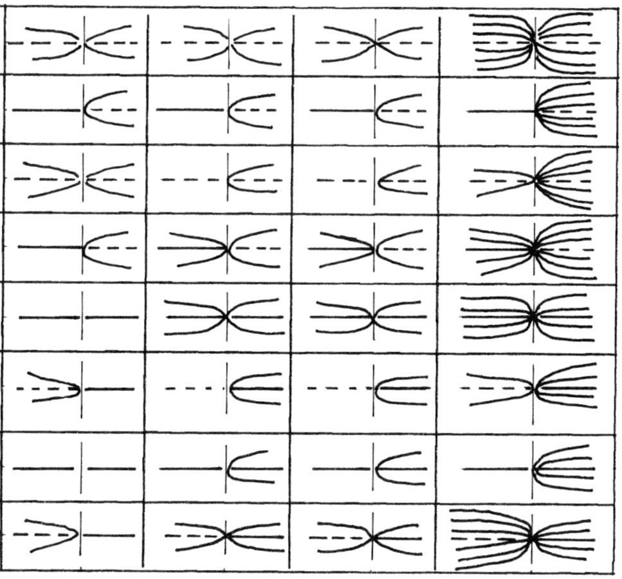

table VII.3, $b_1.b_3 < 0$, $b_2.b_3 > 0$, $b_1.b_2 < 0$

	μ_2	$\bar{\mu}_3$	μ_1	$\bar{\mu}_2$	$\bar{\mu}_1$	branching diagram of segment 3	segment 2	segment 1	all segments
$b_1 > o$, $b_2 < o$	e	e	e	e	e				
		o	o	o	o				
	o	e	e	o	o				
		o	o	e	e				
$b_1 < o$, $b_2 > o$	e	e	e	e	e				
		o	o	o	o				
	o	e	e	o	o				
		o	o	e	e				

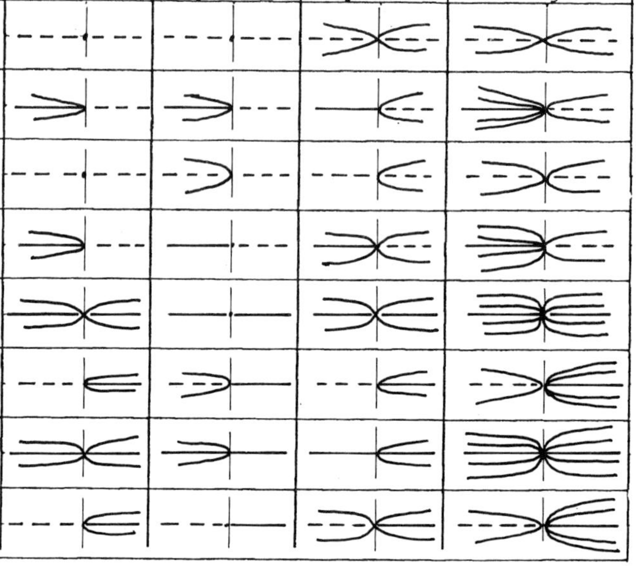

table VII.4, $b_1.b_3 < 0$, $b_2.b_3 < 0$, $b_1.b_2 > o$

$b_1 > o$, $b_2 > o$	e	e	e	e	e				
		o	o	o	o				
	o	e	e	o	o				
		o	o	e	e				
$b_1 < o$, $b_2 < o$	e	e	e	e	e				
		o	o	o	o				
	o	e	e	o	o				
		o	o	e	e				

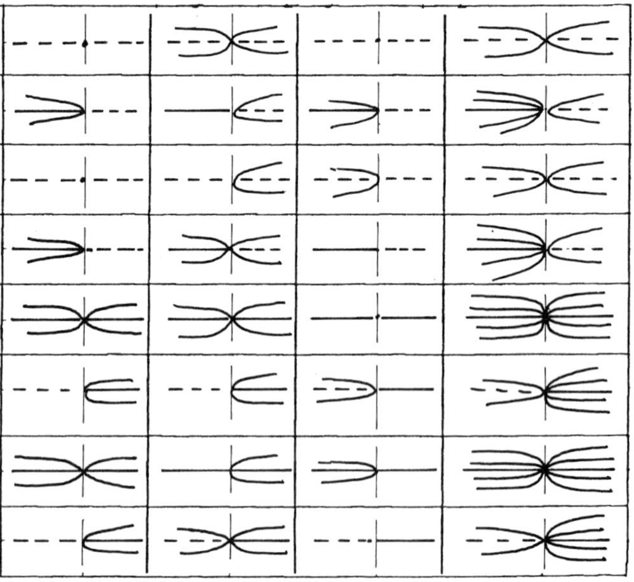

The branching diagrams of table VII.1 are taken from column 2 of table IIIb.2 , row 1 to 4 and row 9 to 12 (segment 3), table IIIb.31, row 1 to 4 and table IIIb.32, row 9 to 12 (segment 2) and table IIIb.1, row 1 to 4 and row 9 to 12 (segment 1).

The branching diagrams of table VII.2 are taken from column 2 of table IIIb.2 , row 1 to 4 and row 9 to 12 (segment 3), table IIIa.31, row 9 to 12 and table IIIa.32 row 1 to 4 (segment 2) and table IIIa.1, row 1 to 4 and row 9 to 12 (segment 1).

The branching diagrams of table VII.3 are taken from column 2 of table IIIa.2 , row 1 to 4 and row 9 to 12 (segment 3), table IIIb.31, row 9 to 12 and table IIIb.32, row 1 to 4 (segment 2) and table IIIa.1, row 1 to 4 and row 9 to 12 (segment 1).

The branching diagrams of table VII.4 are taken from column 2 of table IIIa.2 , row 1 to 4 and row 9 to 12 (segment 3), table IIIa.31, row 1 to 4 and table IIIa.32, row 9 to 12 (segment 2) and table IIIb.1, row 1 to 4 and row 9 to 12 (segment 1).

All branching diagrams of table VII.1 to table VII.4 satisfy the embedding properties of theorem (7) which correspond to an asymptotical stable trivial solution of (1).

Applying the same procedure to the cases r_1 and/or r_2 even, it is easily seen that the embedding properties of theorem (7) are not always satisfied for an asymptotical stable trivial solution of (1). Therefore these cases have to be excluded from theorem (7).

The various branching diagrams corresponding to an unstable trivial solution of (1) are treated in analogy to the asymptotical stable case.

It is obvious by inspection of the different tables III that a combination of the branching diagrams of section 1,2 and 3 corresponding to m even and r_2 odd (r_1 even or odd) cannot produce embedding diagrams of theorem (7) -this statement does not hold for r_2 even- .

The remaining cases which have to be investigated are those corresponding to m, r_1 and r_2 odd and $g < o$. If they too cannot produce embedding-diagrams of the type of theorem (7), then this theorem is proved for $l = 3$. These cases are collected in tables VIII.1 to VIII.4.

The branching diagrams of table VIII.1 are taken from column 6 of table IIIb.2 , row 1 to 4 and row 9 to 12 (segment 3), table IIIb.31, row 1 to 4 and table IIIb.32, row 9 to 12 (segment 2) and table IIIb.1, row 1 to 4 and row 9 to 12 (segment 1).

The branching diagrams of table VIII.2 are taken from column 6 of table IIIb.2 , row 9 to 12 and row 1 to 4 (segment 3), table IIIa.31, row 9 to 12 and table IIIa.32, row 1 to 4 (segment 2) and table IIIa.1, row 1 to 4 and row 9 to 12 (segment 1).

The branching diagrams of table VIII.3 are taken from column 6 of table IIIa.2 , row 9 to 12 and row 1 to 4 (segment 3), table IIIb.31, row 9 to 12 and table IIIb.32, row 1 to 4 (segment 2) and table IIIa.1, row 1 to 4 and row 9 to 12 (segment 1).

The branching diagrams of table VIII.4 are taken from column 6 of table IIIa.2 , row 9 to 12 and row 1 to 4 (segment 3), table IIIa.31, row 1 to 4 and table IIIa.32, row 9 to 12 (segment 2) and table IIIb.1, row 1 to 4 and row 9 to 12 (segment 1).

None of the branching diagrams of table VIII.1 to VIII.4 satisfies the embedding properties of theorem (7) for an asymptotical stable solution of (1).

This is proved in theorem (7) for $l = 3$.

The preceding case ($l=3$) demonstrates the generating rules for $l > 3$, l finite. In the general case $l > o$, the same tables III can be used with some indices interchanged as follows:

1. segment: all indices of table IIIa.1 and table IIIb.1 remain unchanged.

i. intermediate segment: the indices 2 resp. 3 are replaced by i resp. by i+1 , $1 < i < l$.

l. segment: the index 2 is replaced by l.

The allowed branching diagrams are selected (in complete analogy to the cases discussed above) by using the tables II, III and VII and by applying the equations (16), (17.1) and an obvious extension of table VI.

Remark 9:

Theorem (7) is a generalization of theorem (6). It is of importance as in general a l.p.p.r.(1) system (7) will lead to a corresponding n.p.p.r.(1) system (13') or even to a m.p.p.r.(1) system (13'').

The preceding theorems may be illustrated by the examples 1,2 and 3 in (H.Hahn, 1977). It is easily seen that in example 1 and 2 the trivial solution of the critical system ($k = o$) is unstable, whereas

table VIII.1, $b_1.b_3 > 0$, $b_2.b_3 > 0$, $b_1.b_2 > 0$

	μ_2	$\bar{\mu}_3$	μ_1	$\bar{\mu}_2$	$\bar{\mu}_1$
$b_1 > o$, $b_2 > o$	e	e	e	e	e
		o	o	o	o
	o	e	e	o	o
		o	o	e	e
$b_1 < o$, $b_2 < o$	e	e	e	e	e
		o	o	o	o
	o	e	e	o	o
		o	o	e	e

branching diagram of: segment 3 | segment 2 | segment 1 | all segments

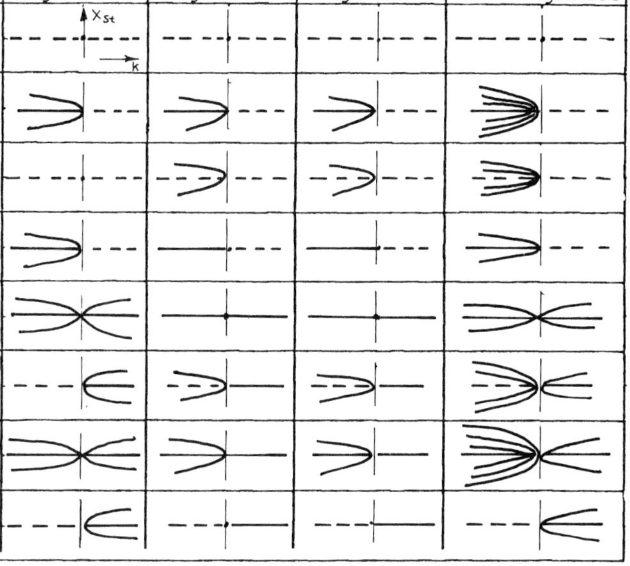

table VIII.2, $b_1.b_3 > 0$, $b_2.b_3 < 0$, $b_1.b_2 < 0$

	μ_2	$\bar{\mu}_3$	μ_1	$\bar{\mu}_2$	$\bar{\mu}_1$
$b_1 > 0$, $b_2 < 0$	e	e	e	e	e
		o	o	o	o
	o	e	e	o	o
		o	o	e	e
$b_1 < 0$, $b_2 > 0$	e	e	e	e	e
		o	o	o	o
	o	e	e	o	o
		o	o	e	e

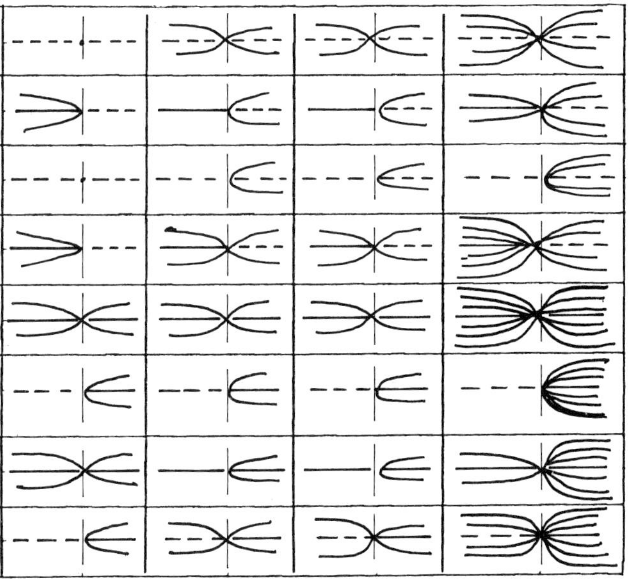

table VIII.3, $b_1.b_3 < 0$, $b_2.b_3 > 0$, $b_1.b_2 < 0$

	μ_2	$\bar{\mu}_3$	μ_1	$\bar{\mu}_2$	$\bar{\mu}_1$
$b_1 > o$, $b_2 < o$	e	e	e	e	e
		o	o	o	o
	o	e	e	o	o
		o	o	e	e
$b_1 < o$, $b_2 > o$	e	e	e	e	e
		o	o	o	o
	o	e	e	o	o
		o	o	e	e

branching diagram of: segment 3 | segment 2 | segment 1 | all segments

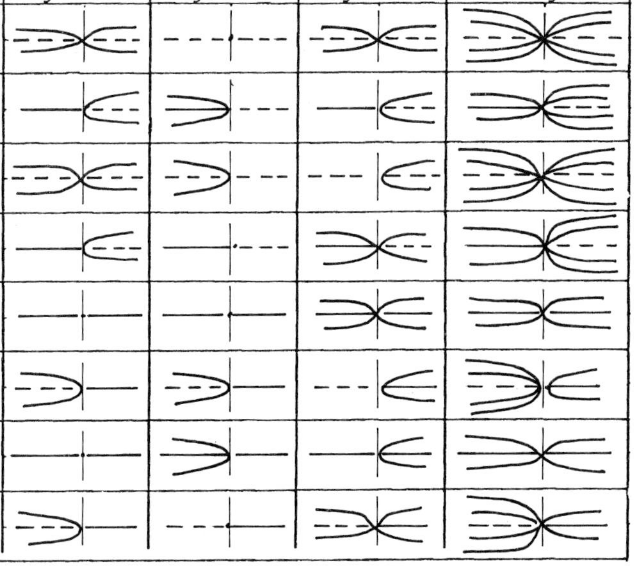

table VIII.4, $b_1.b_3 < 0$, $b_2.b_3 < 0$, $b_1.b_2 > 0$

	μ_2	$\bar{\mu}_3$	μ_1	$\bar{\mu}_2$	$\bar{\mu}_1$
$b_1 > o$, $b_2 > o$	e	e	e	e	e
		o	o	o	o
	o	e	e	o	o
		o	o	e	e
$b_1 < o$, $b_2 < o$	e	e	e	e	e
		o	o	o	o
	o	e	e	o	o
		o	o	e	e

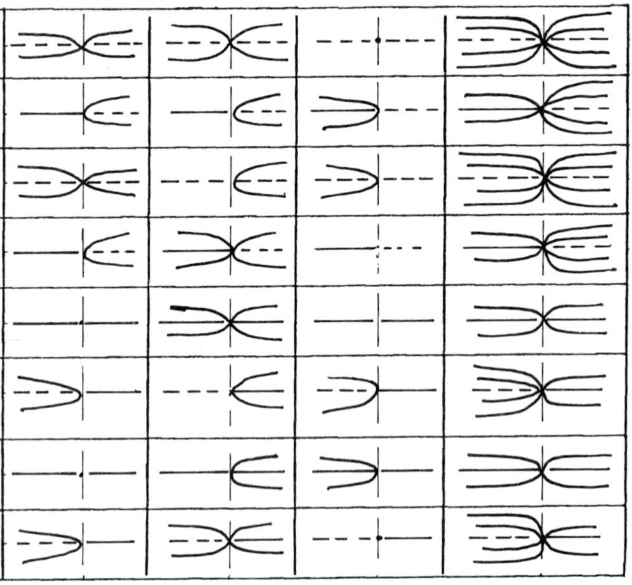

the trivial solution of the critical system of example 3 is asymptotically stable. In all of these examples both,the perturbed system (7) and the corresponding branching equation (13) are l.p.p.r. (1) systems.

Final observation:

The preceding investigations demonstrate, that the stability of a class of critical nonlinear systems may be decided by means of a discussion of the branching equation of a related linear or nonlinear parametrically perturbed system.
This approach delivers a deeper understanding of critical nonlinear systems and demonstrates, that the stability behaviour of a class of these systems may be decided by means of purely algebraic methods (linear eigenvalue theory in combination with a discussion of the algebraic branching equation of the system). Moreover these results have a strong geometric appeal.
These results turn out to be useful in the classification of problems in connexion with a sensitivity analysis of nonlinear systems as well as in special problems of singular perturbation theory.
For the sake of clarity and brevity the corresponding statements have been restricted to a simple critical eigenvalue of the reduced system (1).
On additional assumptions similar results hold for general critical systems with more then one critical eigenvalue and with complex conjugate eigenvalues, having vanishing real part. These results are stated (in a yet vague form) in terms of the following conjecture, which will be stated more precise and proved in a later paper.

CONJECTURE:

A critical nonlinear system $(1^{o})^{5*}$ is asymptotically stable iff there exists a parametrically perturbed related system of branching equations (13°) of dimension s (s= number of the critical eigenvalues of the reduced system (1°), $1 \leqslant s$), where $r_1, \ldots, r_l$ are odd and (13°) satisfies an additional "condition A" such that

a) the asymptotical stable segments of the trivial solution of (7°) are smoothly embedded by means of 4.n "branching-manifolds" (n= o,1,2, ...) and

b) the unstable segments of the trivial solution of (7°) are smoothly embedded by 2.(1+2.n) branching manifolds (n= o,1, ...), where in case of complex conjugate critical eigenvalues, i.g. in case of a Hopf-Bifurcation, each periodic solution is counted twice (as an integralmanifold).

(5[*] The index "o" takes into account, that the assumption 1 of equation (1) is no longer needed, and that the branching equation (13) resp. (13°) may represent a system of equations of dimension s , $s \geqslant 1$).

" Condition A" represents a set of additional assumptions which may be (informally) interpreted as follows:

From an algebraic point of view this condition implies (among others), that equation (13°) does not have vanishing resultants resp.vanishing intermediate resultants.
Moreover this assumption contains (because of technical reasons) some conditions with respect to the allowed form of the Newton-Diagram of the branching equation (compare theorem 6).

Abstract-Theorems are proved which admit a stability decision for nonlinear critical systems in terms of existence criteria for bifurcating stationary solutions of a "parametrically perturbed related system". The criteria of the main theorem are pure algebraic and have a strong appeal to geometry. The concept of "smoth embedding of a stationary solution" as well as a specific technique from bifurcation theory are used to prove the theorems. A generalization of these theorems is presented in form of a conjecture.

Acknowledgment

The author is most grateful to Prof. Dr. W. Hahn, Mathemat. Institut II der Techn. Universität Graz, for stimulating discussions on the problem of safe resp. of dangerous segments of the stability boundary of dynamical systems (compare Bautin's paper) and to Prof. Dr. J. George, Math. Department of the University of Wyoming, Laramie, for useful discussions on the application of Newton's Diagram in bifurcation theory.
This work has been supported by the Deutsche Forschungsgemeinschaft.

References

1. N. N. Bautin, Das Verhalten dynamischer Systeme an den Grenzen des Stabilitätsbereichs, Moskau (195o), (russ.).
2. W. Forrester, World Dynamics, Cambridge, Wright-Allen Press (1971).
3. H. Hahn, Zur Diskrepanz zwischen Theorie und Praxis nichtlinearer Regelkreise, Preprint Universität Tübingen (1975).

4. H. Hahn, Zum Problem der praktischen Stabilität, Berichte der Mathematisch-Statistischen Sektion im Forschungszentrum Graz, (1976).

5. H. Hahn, The application of root-locus-technique to nonlinear control systems with multiple steady states, Intern. J. of Control, (1977).

6. H. Hahn, Theorie und Anwendung des "Newton-Diagramms" bei nichtlinearen Regelsystemen, Preprint Universität Tübingen,(1976).

7. W. Hahn, Stability of Motion, Springer Verlag, (1967).

8. J. Hale, J. Diff. Eq. 1o (1971),PP.59 - 82.

9. R. Hausrath , J. Diff. Eq. 13 (1973), pp. 329 - 357.

1o. J. Kelly, J. Math. Anal. Appl. (1967), pp. 336 - 344.

11. J.E. Marsden, M. Mc Cracken, The Hopf Bifurcation and its Application, Springer (1976).

12. D. Meadows et al , The Limits to Growth, Universe Books, (1972).

13. M. Mesarovic, E. Pestel, A Goal-seeking and Regionalized Model for Analysis of Critical World-Relationship - A Conceptional Foundation, Kybernetes , (1972).

14. W. Reissig, G. Sansone, R. Conti, Nichtlineare Differentialgleichungen höherer Ordnung, Cremonese, (1969).

15. M.M. Vainberg, V. A. Trenogin, Theory of branching of solutions of nonlinear equations, Noordhoff, (1974).

ASYMPTOTIC SYSTEM ANALYSIS AND GOAL FUNCTIONS

G. Palm
Max-Planck-Institut für biologische Kybernetik
Spemannstrasse 38
74 Tübingen

0. Introduction

The aim of this paper is to provide a tool for an analysis of the asymptotic behaviour (in short: asymptotic analysis) of dynamical systems. I shall use a discrete time, state space description of the system (cf. section 1). Let us assume that some properties of the asymptotic behaviour of a particular system are expected on heuristic grounds - a situation that may arise in two different ways:

1) The behaviour of the system has been observed thoroughly for a long time,
2) The system has been designed to achieve a certain goal asymptotically.

Then one may try to translate these heuristic ideas into a mathematical language in order to use them as an additional tool in a general asymptotic analysis of the system, or to prove that the system will indeed behave asymptotically as expected.
In this situation I want to suggest the following: One should try to express heuristic ideas on the system's asymptotic behaviour in terms of a goal function g, defined on the state space, the values of which are always decreasing during the evolution of the system in time.
In this way one can express a tendency of the system to move towards the region C in state space where g is small. This idea is very similar to the idea underlying the concept of a Liapunov function. The main difference is the following: For a Liapunov function the region C usually consists of only one (invariant) point. Therefore a Liapunov function can often be only defined in a small neighbourhood of this point (most systems have more than one invariant point); sometimes additional conditions are imposed, which guarantee that

the system will asymptotically move into just that point (see for example [4] or [17], p. 121). In this paper, however, I shall consider functions defined globally, i.e. on the whole state space, and I shall characterize the region C (usually not containing only one point) into which the system tends to move.

A broad experience with Liapunov functions illustrates that it is still very hard to define such a function globally (even if the system has only one invariant point, and this point is stable). Therefore I have substantially weakened the requirement that the values of g should always decrease during time evolution (compare Def. 1.0. and 1.1.); even so it is still possible to show that the system asymptotically moves into the region C.

In the first section the basic ideas and results are formulated. The second section discusses possible applications.

1. Goal functions

We assume that the time evolution of a system is given explicitly in the following way:

The state of a system is completely described by a point x in a state space X. The time evolution is given by a mapping $\varphi: \begin{cases} X \to X \\ x \to \varphi(x) \end{cases}$ which associates with a state x (say at time t) the next state $\varphi(x)$ (in which the system will be at time t+1).

In the following we will consider real valued functions f,g,.. on X, and as usual we will write

$f=g$ iff $f(x)=g(x)$ for every $x \in X$,

$f \leqq g$ iff $f(x) \leqq g(x)$ for every $x \in X$, and

$f<g$ iff $f \leqq g$ and $f \neq g$.

The following definition is used to present the essential ideas in the final concept of a goal function (Def.1.1).

1.0. Definition: A bounded function $f: X \to \mathbb{R}$ is called a global L-function, if $f \circ \varphi < f$.

Let me explain the meaning of this definition: $f \circ \varphi \leqq f$ means

1) $f(\varphi(x)) = f \circ \varphi(x) \leqq f(x)$ for every $x \in X$, and

2) there is an $x \in X$ such that $f(\varphi(x)) < f(x)$.

Taking inequality (1) at the point $\varphi(x) \in X$ we find that $f(\varphi(\varphi(x))) \leqq f(\varphi(x))$. Then taking (1) at $\varphi(\varphi(x)) = \varphi^2(x)$, $\varphi^3(x)$ and so forth, we get

3) $f(x) \geq f(\varphi(x)) \geqq f(\varphi^2(x)) \geqq f(\varphi^3(x)) \geqq \dots$.

In other words: the point x moves always into the direction of lower f-values.
Since the sequence (3) is monotonically decreasing and bounded (for every $t \in \mathbb{N}$ is $f(\varphi^t(x)) \geqq \inf \{f(z) : z \in X\}$), it converges and we may call the limit $f^*(x)$ and note that

4) $f \geqq f \circ \varphi^t \geqq f^*$ for every $t \in \mathbb{N}$.

Thus we see that, starting at the point $x \in X$, for increasing time t the system tends to move into that region C_x of the state space, where f equals $f^*(x)$:

5) $C_x = \{z \in X : f(z) = f^*(x)\}$.

I will weaken the requirement (3) in the following sense: It may happen that $f(\varphi(x)) > f(x)$, but this so rarely (and with so small values for $f(\varphi(x)) - f(x)$) that, averaging over sufficiently large times, the point x tends to move into the direction of lower f - values.
To formulate this precisely let me introduce some additional notation: It will be convenient to consider a (sufficiently rich) set $\mathcal{E}$ of functions $f: X \to \mathbb{R}$ (for example we could take $\mathcal{E}$ = the set of <u>all</u> functions $f: X \to \mathbb{R}$) and express the time evolution φ by the corresponding operator $T: \begin{cases} \mathcal{E} \to \mathcal{E} \\ f \to f \circ \varphi \end{cases}$. Then we consider the powers T^t $(t \in \mathbb{N})$ of T (T^0 being defined as the identity mapping on $\mathcal{E}$) and their means $T_n f := \frac{1}{n} \sum_{i=0}^{n-1} T^i f$. $T_\infty f$ is defined by $T_\infty f(x) := \lim_n T_n f(x)$, if this limit exists for every $x \in X$.

<u>1.1. Definition</u>: A function $g \in \mathcal{E}$ is called a <u>goal function</u>, if $T_\infty g < g$.

<u>1.2. Proposition</u>: Every global L - function $f \in \mathcal{E}$ is a goal function.

<u>Proof</u>: $T^i f = T^{i-1} Tf = T^{i-1}(f \circ \varphi) = T^{i-2}(T(f \circ \varphi)) = T^{i-2}(f \circ \varphi^2) = T^{i-3}(f \circ \varphi^3) = \ldots$
$= T(f \circ \varphi^{i-1}) = f \circ \varphi^i$.

Thus (3) and the boundedness of f imply that the sequence $T^i f(x) = f \circ \varphi^i(x) = f(\varphi^i(x))$ $(i \in \mathbb{N})$ converges for every $x \in X$. A fortiory the sequence $T_n f(x)$ $(n \in \mathbb{N})$ converges to the same limit, and we get $T_\infty f = f^* < f$ by (4).

The next step is to show that the existence of a goal function still implies that, for increasing time t, the system tends to move into a certain region of the state space X - like in (5). Such a statement is in contradiction with ergodic properties of the system. For example the existence of a finite φ- invariant measure on X implies (by Poincaré's Recurrence Theorem) that for almost every starting point x the system will eventually come back close to x; and this would mean that the region of the state space to which the system tends to move has to contain almost every point x of X. In this case it would be nonsense to speak of a tendency to move into that region. Proposition 1.3 restates this contradiction: if one describes a system using a state space X, one has to decide between having a motion invariant measure on X, or a motion decreasing (resp. motion increasing) function on X. The implications of proposition 1.3 in a thermodynamical context are discussed in section 2.

<u>1.3. Proposition:</u> Let p be a finite φ-invariant measure on X. Take $\mathcal{E} = L^1(X,p)$ the space of all real valued functions f on X, such that $|f|$ is integrable with respect to p. Let $g \in \mathcal{E}$ be a goal function. Then $p(\{z \in X: g(z) > T_\infty g(z)\}) = 0$. I.e. for p-almost every $x \in X$ $g(x)$ equals $T_\infty g(x)$.

<u>Proof:</u> Since p is φ-invariant $\int f dp = \int Tf dp = \int T^i f dp$ for every $i \in \mathbb{N}$ and every $f \in L^1(X,p)$. Therefore $\int (g-T_n g)\, dp = \frac{1}{n} \sum_{i=0}^{n-1} \int (g-T^i g)\, dp = 0$ for $n \in \mathbb{N}$ and by Fatou's Lemma $\int (g-T_\infty g)\, dp = 0$.
Since g is a goal function $g-T_\infty g \geqq 0$, and thus $g-T_\infty g = 0$ p-almost everywhere.

<u>1.4. Definition:</u> Let p be a measure on X. A goal function $g \in \mathcal{E} = L^1(X,p)$ is called a <u>goal function for</u> p, if $p(\{z \in X: g(z) > T_\infty g(z)\}) \neq 0$.

Using this definition we can reformulate proposition 1.3 :

<u>1.5. Proposition:</u> There is no goal function for a finite φ-invariant measure on X.

Finally I will characterize more explicitely the subsets of X into which a system tends to move, provided it has a goal function g. To this end I will use the following intuitive language: $\varphi^t(x) \in M \subseteq X$ is expressed by saying: starting in x, M is visited at time t.
If the limit exists, $h(x,M) := \lim_n \frac{1}{n} \sum_{i=0}^{n-1} 1_{[\varphi^i(x) \in M]}$ expresses the frequency of visits in M starting in X.
For $x \in X$ and $\epsilon > 0$ let $A_x := \{z \in X: g(z) < T_\infty g(x)\}$,
$A_{x,\epsilon} := \{z \in X: g(z) \geqq T_\infty g(x) + \epsilon\}$, $A_\epsilon := \{z \in X: g(z) \geqq T_\infty g(z) + \epsilon\}$,
$C_{x,\epsilon} := X \setminus (A_x \cup A_{x,\epsilon}) = \{z \in X: T_\infty g(x) \leqq g(z) < T_\infty g(x) + \epsilon\}$,
$C_\epsilon := X \setminus A_\epsilon = \{z \in X: T_\infty g(z) \leqq g(z) < T_\infty g(z) + \epsilon\}$,
$C_x := \bigcap_{\epsilon>0} C_{x,\epsilon} = \{z \in X: g(z) = T_\infty g(x)\}$, $C := \bigcap_{\epsilon>0} C_\epsilon = \{z \in X: g(z) = T_\infty g(z)\}$.

<u>1.6. Proposition:</u> Let g be a global L-function. Starting in $x \in X$, A_x is never visited and $A_{x,\epsilon}$ as well as A_ϵ only a finite number of times for every $\epsilon > 0$.

Proof: 1) A_x is never visited:

$\varphi^n(x) \in A_x$ means $T^n g(x) < T_\infty g(x)$, but this is impossible since $g(x) \geq T_\infty g(x)$ implies $T^n g(x) \geq T^n T_\infty g(x) = T_\infty g(x)$.

2) $A_{x,\varepsilon}$ is visited a finite number of times:

Suppose $A_{x,\varepsilon}$ was visited infinitely often, say at times t_i $(i \in \mathbb{N})$. Then $T^{t_i} g(x) \geq T_\infty g(x) + \varepsilon$. From the proof of proposition 1.2 we know that $T_\infty g(x) = \lim_n T^n g(x) = \lim_i T^{t_i} g(x)$, yielding a contradiction.

3) A_ε is visited a finite number of times:

Since $T_\infty g(x) = T_\infty T^n g(x) = T_\infty g(\varphi^n(x))$, we have $\varphi^n(x) \in A_{x,\varepsilon}$ iff $\varphi^n(x) \in A_\varepsilon$.

1.7. Proposition: Let g be a goal function. Starting in $x \in X$, A_x is never visited and $h(x, A_{x,\varepsilon}) = h(x, A_\varepsilon) = 0$ for every $\varepsilon > 0$. Thus $h(x, C_{x,\varepsilon}) = h(x, C_\varepsilon) = 1$ for every $\varepsilon > 0$.

Proof: 1) A_x is never visited: see (1) in the proof of 1.6.

2) $h(x, A_{x,\varepsilon}) = 0$: Suppose $h(x, A_{x,\varepsilon}) \neq 0$ or the limit does not exist. In any case there is a constant $c > 0$ and a sequence t_i $(i \in \mathbb{N})$ of integers such that $t_i^{-1} \cdot \sum_{k=0}^{t_i-1} 1_{[\varphi^k(x) \in A_{x,\varepsilon}]} \geq c$.

We know that $g(\varphi^k(x)) \geq T_\infty g(x)$ for every $k \in \mathbb{N}$ and $g(\varphi^k(x)) \geq T_\infty g(x) + \varepsilon$, if $\varphi^k(x) \in A_{x,\varepsilon}$. Therefore

$$T_{t_i} g(x) = t_i^{-1} \cdot \sum_{k=0}^{t_i-1} g(\varphi^k(x))$$
$$\geq t_i^{-1} \cdot \Sigma_k (1_{[\varphi^k(x) \in A_{x,\varepsilon}]} \cdot (T_\infty g(x) + \varepsilon) + 1_{[\varphi^k(x) \notin A_{x,\varepsilon}]} \cdot T_\infty g(x))$$
$$= T_\infty g(x) + \varepsilon \cdot t_i^{-1} \cdot \Sigma_k 1_{[\varphi^k(x) \in A_{x,\varepsilon}]} \geq T_\infty g(x) + c \cdot \varepsilon,$$

in contradiction to $T_\infty g(x) = \lim_i T_{t_i} g(x)$.

3) $h(x, A_\varepsilon) = 0$: see (3) in the proof of 1.6.

One has to be careful in the interpretation of this proposition, since the sets C and C_x, defined above, may be empty.

1.8. Example: Let $X = \mathbb{R}$, $\varphi: \begin{cases} X \to X \\ x \to x+1 \end{cases}$, then $g(x) = \operatorname{artan} x$ is a goal function (it is even a global L-function). In this example $T_\infty g(x) = \frac{\pi}{2}$ for every $x \in X$ and therefore $C_x = C = \emptyset$.

However, if X is compact and φ continuous this can not occur.

1.9. Proposition: Let X be compact, $\varphi: X \to X$ continuous, $\mathcal{E} = C(X)$, $g \in \mathcal{E}$ a goal function. Then $C \neq \emptyset$ and $C_x \neq \emptyset$ for every $x \in X$. Moreover $T_\infty g(X) = T_\infty g(C) = g(C)$.

<u>Proof</u>: Clearly the assertions of proposition 1.7 again hold.

1) C and C_x are nonempty since they are intersections of directed families of nonempty closed sets (C_ε and $C_{x,\varepsilon}$ respectively).

2) $T_\infty g(X) \subseteq g(C)$: Let $x \in X$. We have to find a $z \in C$ such that $g(z) = T_\infty g(x)$. Let $n \in \mathbb{N}$, then $h(x, C_{1/n}) = 1$ implies that there is a $t_n \in \mathbb{N}$ satisfying $T_\infty g(x) \leq g(\varphi^{t_n}(x)) \leq T_\infty g(x) + 1/n$.
Thus we get a sequence t_n with $g(\varphi^{t_n}(x)) \to T_\infty g(x)$. If z is a cluster point of the sequence $\varphi^{t_n}(x)$ in X, then the continuity of g implies $g(z) = T_\infty g(x)$.

3) Clearly $g(C) = T_\infty g(C) \subseteq T_\infty g(X)$.

For propositions 1.3. and 1.5. we did not need the fact that the operator T on was given by a point-to-point transformation φ of X. A similar generalization of propositions 1.6. and 1.7. requires some more effort. Since it will be important for statistical and thermodynamical applications this generalization is carried out in appendix A.

2. Possible applications

The second principle of thermodynamics states that the Boltzmann-entropy S of a closed physical system is nonincreasing during time evolution.
Since the entropy can be regarded as a function on the Liouville space characterizing the system, this statement means that S is a global L-function. However, many textbooks on thermodynamics or statistical mechanics (e.g. [7], [9]) remark that this law does not hold strictly, but only in the mean, over a sufficiently large time (which is argued to be quite short in practice). In other words: entropy may be only a goal function.
As we saw in Prop. 1.3. and 1.5. this weakened requirement on the time evolution of physical entropy is still in disagreement with the existence of an invariant measure on the phase space (like the Liouville measure for closed systems), if entropy is regarded as a function on the phase space. Therefore, in thermodynamics entropy is usually defined on a different space - namely the Liouville space.
The basic idea is the following: In many systems (X, φ), which admit a φ-invariant measure on X and are ergodic, there is a way of expressing the ergodicity of (X, φ) (which practically means that there is no tendency in the asymptotic behaviour) by means of a goal function on a different state-space model (X', φ') describing basically the same phenomena: Take a space X' of probability

measures on X (if X is the usual physical phase space, X' corresponds to the Liouville space). Define $\varphi'(p)$ for $p \in X'$ by $\varphi'(p)(A) := p(\varphi^{-1}(A))$ for every measurable set $A \subseteq X$. Then the ergodicity of X can often be described by the fact that the system (X', φ') tends to move towards a single φ-invariant measure $\bar{p}$ on X, and this tendency can be expressed by means of a goal function (see Appendix B). If $\bar{p}$ is the uniform distribution on X, the so-called coarse grain entropy (cf. [5]) can be taken as a goal function. One has to be aware of the fact that the systems (X, φ) and (X', φ') generally display a different dynamical behaviour. In building a state-space model of a class of physical phenomena one still has to decide between a model admitting an invariant measure (like (X, φ)) and a model admitting a goal function (like (X', φ')). In principle this decision should be made according to the phenomenology: one should choose that model that fits the observed data better.
However, our measuring techniques are usually incapable of deciding between these two alternatives (for example Poincaré cycles are usually very long), so that this decision essentially remains a problem of taste.
Let us move from these more theoretical considerations to problems encountered in concrete mathematical models of a known phenomenology. In many cases one wants a model, that expresses a certain tendency in the time evolution. Accordingly one may look for Liapunov functions or global L - functions to express this tendency. In certain systems of differential equations (that often arise from chemical considerations) Prigogine, Schlögl and others ([2], [3], [14], [15], [16]) have proposed to use as Liapunov functions certain functions that are closely related to entropy (for an overview see [11]). In concrete examples, however, these functions usually turn out not to be global L - functions. I conjecture that some of them may be goal functions.
Let me illustrate this conjecture by considering in some more detail a model of Schlögl [14], who has proposed the information gain (a concept derived from information theory) as a Liapunov function. The connection between Schlögl's idea and the Glansdorf-Prigogine criterion is illustrated in [15] and [11].
In Schlögl's model the statistical state of a system is given by a probability vector $p = (p_1, \ldots, p_n)$, i.e.
$X = \{p = (p_1,\ldots,p_n) : \Sigma_i \, p_i = 1, \; p_i \geq 0$ for every $i=1,\ldots,n\}$, and the time evolution is given by a transition matrix $Q: \begin{cases} X \to X \\ (p_j)_j \to (\Sigma_i \, p_i Q_{ij})_j \end{cases}$.

In other words: the time evolution is described as a Markov process of 'memorylength' 1. There is then a probability vector $\bar{p}$ such that $\bar{p} = Q\bar{p}$ and we may consider the function $g:\begin{cases} X \to \mathbb{R} \\ p \to I(p|\bar{p}) \end{cases}$, where the information gain (or missing information [10]) is defined by
$I(p|\bar{p}) := \Sigma_i\, p_i \log(p_i/\bar{p}_i)$.

2.1. Proposition (Schlögl): $g \geq 0$ is a global L - function.

Proof: Note that $\log x \leq x - 1$ for every $x>0$.

1) $g \geq 0$: $-I(p|\bar{p}) = \Sigma_j\, p_j \log(\bar{p}_j/p_j) \leq \Sigma_j\, p_j((\bar{p}_j/p_j)-1) = 0$.

2) $g \circ Q \leq g$, i.e. $I(Qp|\bar{p}) \leq I(p|\bar{p})$ for every $p \in X$:

$$I(Qp|\bar{p}) - I(p|\bar{p}) = \Sigma_j (Qp)_j \cdot \log((Qp)_j/\bar{p}_j) - \Sigma_i\, p_i \log(p_i/\bar{p}_i) =$$
$$= \Sigma_j (\Sigma_i\, p_i Q_{ij}) \log((Qp)_j/\bar{p}_j) - \Sigma_i\, p_i (\Sigma_j\, Q_{ij}) \log(p_i/\bar{p}_i) =$$
$$= \Sigma_{ij}\, p_i Q_{ij} \cdot \log((Qp)_j\, \bar{p}_i/(p_i\, \bar{p}_j))$$
$$\leq \Sigma_{ij}\, p_i Q_{ij}((Qp)_j\, \bar{p}_i/(p_i\, \bar{p}_j) - 1) =$$
$$= \Sigma_j (\Sigma_i\, \bar{p}_i Q_{ij})(Qp)_j/\bar{p}_j - \Sigma_{ij}\, p_i Q_{ij} = \Sigma_j (Qp)_j - 1 = 0.$$

3) In (2) equality holds iff $(Qp)_j\, \bar{p}_i/(\bar{p}_j\, p_i) = 1$ for all i,j, which implies $p_i/\bar{p}_i = 1$ for all i. Thus for $p \neq \bar{p}$ we have $g(Qp)<g(p)$.

This proposition may be regarded as a proof of the second principle of thermodynamics (in it's strict form) for this kind of system. The problem in this approach lies in the 'coarseness' of the state space X: Here the points of the state space usually do not contain all information on the system. Therefore it may be impossible to describe the time evolution by a mapping $\varphi: X \to X$. In other words: the underlying stochastic process may not be a Markov process of memorylength 1. Therefore more general statistical laws for the time evolution are of interest.

2.2. Example: Consider the following Markov process of memorylength 2: $x(t)$ $(t \in \mathbb{N})$ is a family of random variables with possible values 0 and 1. For $0<a<1$ the transition probabilities are

$pr[x(t+2) = 0|x(t+1) = 0,\ x(t) = 0] = a$

$pr[x(t+2) = 0|x(t+1) = 0,\ x(t) = 1] = 1-a$

$pr[x(t+2) = 0|x(t+1) = 1,\ x(t) = 0] = a$

$pr[x(t+2) = 0|\ x(t+1) = 1, x(t) = 1] = 1-a$.

The initial distribution is given by $pr[x(0) = 0, x(1) = 0] = \frac{1}{2}$, $pr[x(0) = 0,\ x(1) = 1] = \frac{1}{2}$ and $pr[x(0) = 1] = 0$.

Then one can compute the probability distributions at time t, $p^t = (p_0^t, p_1^t) = (pr[x(t) = 0], pr[x(t) = 1])$, which converge to $\bar{p} = (1/2, 1/2)$.

Take $X := \{p = (p_0, p_1): p_0 + p_1 = 1,\ p_0 \geq 0,\ p_1 \geq 0\}$ as the state space. The time evolution (starting from $p^0 = (1,0)$ at time 0 and $p^1 = (\frac{1}{2}, \frac{1}{2})$

at time 1) is given by the probability vectors p^t. For $a \neq \frac{1}{2}$ the graph of $g(p^t) := I(p^t|\bar{p})$ is shown in Figure 1.

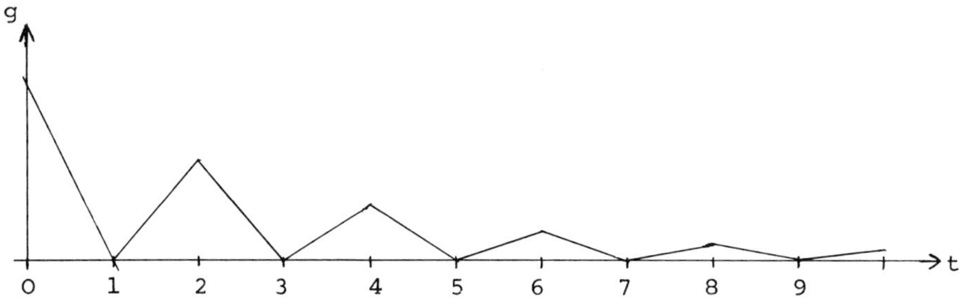

Figure 1

We see that the oszillations of $g(p^t)$, occuring in this example, can be smoothed out by averaging over only two time steps. Accordingly, for Markov processes of memorylength n averaging over n time steps will in general be necessary. It may turn out to be also sufficient. In particular I conjecture that for any Markov process with finite memorylength $g(p^t)$ always has the properties of a goal function. This should still hold, if the memorylength is infinite, provided that the memory is asymptotically decreasing in an appropriate way. The following simple result may illustrate this.

<u>2.3. Proposition</u>: Consider a discrete time stochastic process, i.e. a family $x(t)$ $(t \in \mathbb{N})$ of random variables, with possible values in $\Omega = \{1,\dots,n\}$. Let $p^t = (pr[x(t)=1],\dots,pr[x(t)=n])$ and suppose that $\lim p^t = \bar{p}$ exists. Let $g(p) := I(p|\bar{p})$, then

a) $g(p^t) \geq 0$ for every $t \in \mathbb{N}$,

b) $\lim_t g(p^t) = 0$,

and therefore $g(p^t) \geqq \lim_n n^{-1} \sum_{i=0}^{n-1} g(p^i)$ for every $t \in \mathbb{N}$.

<u>Proof:</u> a) See proposition 2.1.

b) The continuity of g implies $\lim g(p^t) = g(\bar{p}) = 0$.

Having illustrated why the concept of a goal function may be successfully used for an asymptotic analysis of certain systems, I should now mention some concrete applications which are less artificial than example 2.2. One should consider a special state space model that is
1) of practical relevance (as a description of a class of phenomena)
2) sufficiently complex, so that a description by means of a goal function does not seem trivial (in particular, models not admitting a global L - function are of interest.

It would take quite a long time, however, to get enough (heuristic) insight into the behavior of such a system, to be able to propose a goal function for it. Therefore I can only mention one concrete example, namely a cooperative stereo algorithm which I had been discussing with T. Poggio (the analysis is carried out in [8]). But I am sure that there are many specialists having enough insight into certain systems, who may even have built state space models and tried to find Liapunov functions for these, but are not satisfied with the results. I hope that this paper will encourage them to express their heuristic knowledge in terms of a goal function.

Appendix A

The following theorem is a generalization of propositions 1.3 , 1.5 and 1.7. The operator $T:\mathcal{E}\to\mathcal{E}$ is no longer given by a transformation φ of the state space X, but is only assumed to be positive (i.e. $f\geqq 0$ implies $Tf\geqq 0$). The set $\mathcal{E}$ is assumed to be a Banach lattice. The theory of Banach lattices seems to be the natural framework for this kind of generalization (see for example [13] , especially chapters III and V). In analogy to the definitions in section 1 we define

$T_\infty f=\lim_n T_n f=\lim_n n^{-1}\sum_{i=0}^{n-1} T^i f$, and for $p\in E'_+$ and $M\subseteq E'$:

$h(p,M):=\lim_n n^{-1}\sum_{i=0}^{n-1} 1_{[T'^i p\in M]}$.

A1. Proposition: Let E be a Banach lattice, $T:E\to E$ a positive linear operator. Suppose $g\in E$ satisfies $T_\infty g\leqq g$, where $T_\infty g$ exists as a norm limit.

1) $\langle g-T_\infty g, T'_n\mu\rangle\to 0$ for every $\mu\in E'$, and $T_\infty T_\infty g=T_\infty g$.
2) If there is a strictly positive $\mu\in E'$ with $T'\mu=\mu$, then $T_\infty g=g$.
3) For any 'starting distribution' $p\in E'_+$ and any $\epsilon>0$ let
   $C_{p,\epsilon}:=\{\mu\in E'_+ : \langle T_\infty g,p\rangle\leqq\langle g,\mu\rangle\leqq\langle T_\infty g,p\rangle+\epsilon\}$ and
   $C_\epsilon:=\{\mu\in E'_+ : \langle T_\infty g,\mu\rangle\leqq\langle g,\mu\rangle\leqq\langle T_\infty g,\mu\rangle+\epsilon\}$.
   Then $h(p,C_\epsilon)=h(p,C_{p,\epsilon})=1$.
4) For any $p\in E'_+$, any $\epsilon>0$ and any $f\in E$ let $C_{\epsilon,f}:=\{\mu\in E'_+ : \langle f,\mu\rangle\leqq\epsilon\}$
   Then $h(p,C_{\epsilon,f})=1$, if $T_\infty f=0$ as a norm limit.

Proof: $TT_\infty g=T_\infty g$ and therefore $T_nT_\infty g=T_\infty g$ and $T_\infty g=\lim_n T_nT_\infty g=T_\infty T_\infty g$. Thus $\langle g-T_\infty g,T'_n\mu\rangle=\langle T_ng-T_nT_\infty g,\mu\rangle=\langle T_ng-T_\infty g,\mu\rangle\to 0$.
2) $\langle g-T_\infty g,\mu\rangle=\langle g-T_\infty g,T'_n\mu\rangle\to 0$ by (1). Thus $\langle g-T_\infty g,\mu\rangle=0$, which implies $g=T_\infty g$, since μ is strictly positive and $g-T_\infty g\geqq 0$.
3) $\langle T_\infty g,p\rangle=\langle T^nT_\infty g,p\rangle=\langle T_\infty g,T'^np\rangle$. Thus $T'^np\in C_\epsilon$ iff $T'^np\in C_{p,\epsilon}$.

Next define $f:=g-T_\infty g$, then $T_\infty f = T_\infty g - T_\infty T_\infty g = 0$. $T_\infty g \leqq g$ implies $\langle T_\infty g,\mu\rangle \leqq \langle g,\mu\rangle$ for every $\mu \in E'_+$. Therefore $\mu \in C_{\epsilon,f}$, i.e. $\langle g,\mu\rangle - \langle T_\infty g,\mu\rangle = \langle f,\mu\rangle \leqq \epsilon$, means $\langle T_\infty g,\mu\rangle \leqq \langle g,\mu\rangle \leqq \langle T_\infty g,\mu\rangle + \epsilon$, i.e. $\mu \in C_\epsilon$. So (3) is reduced to (4).

4) Let p, ϵ and f be given and suppose $h(p,C_{\epsilon,f}) \neq 1$. Then there is a $c>0$ and a sequence t_i $(i \in \mathbb{N})$ of integers, with $t_i^{-1}\sum_{k=0}^{t_i-1} 1_{[T'^k p \in C_{\epsilon,f}]} \leqq 1-c$, i.e. $t_i^{-1}\Sigma_k\, 1_{[T'^k p \notin C_{\epsilon,f}]} \geqq c$.

Now $\lim_n \langle f,T'_n p\rangle = \lim_n \langle T_n f,p\rangle = \langle T_\infty f,p\rangle = 0$ and therefore $\lim_i \langle f, T'_{t_i} p\rangle = 0$. However

$\langle f,T'_{t_i}p\rangle = t_i^{-1}\Sigma_k \langle f,T'^k p\rangle \geqq t_i^{-1}\Sigma_k \langle f,T'^k p\rangle 1_{[T'^k p \notin C_{\epsilon,f}]} \geqq c\,\epsilon$,

a contradiction.

<u>A2.Theorem</u>: Let E be a Banach lattice admitting a strictly positive linear form $\nu \in E'$, $T:E\to E$ a positive linear operator, which is mean ergodic (compare [13],III,7 , the corresponding projection P being T_∞).

1) The following alternative holds:

<u>Either</u> there is a $g \in E$ satisfying $T_\infty g < g$,

<u>or</u> there is a strictly positive $\mu \in E'$ satisfying $T'\mu = \mu$.

Suppose there is a $g \in E$ satisfying $T_\infty g < g$, then the following statements are true:

2) $f \in E_{g-T_\infty g}$ implies $T_\infty f = 0$ (for any $x \in E$, E_x denotes the closed order - ideal generated by x, cf. [13],II,2).

3) For any 'starting distribution' $p \in E'_+$, $T'_\infty p \in (E_{g-T_\infty g})^{\circ}$.

4) For any $p \in E'_+$, any $\epsilon > 0$ and any $f \in E_{g-T_\infty g}$, $h(p,C_{\epsilon,f}) = 1$ ($C_{\epsilon,f}$ being defined as in A1).

<u>Proof</u>: 1) If both alternatives were true A1(2) would yield a contradiction.

Assume that there is no $g \in E$ with $T_\infty g < g$, i.e. for every $g \in E$ $T_\infty g \leqq g$ implies $T_\infty g = g$. We have to find a strictly positive T'-invariant $\mu \in E'_+$. Take $\mu := T'_\infty \nu$, where ν is strictly positive. Then $T'\nu = \nu$. To prove that μ is strictly positive, take any $g \geqq 0$: $0 = \langle g,\mu\rangle = \langle g,T'_\infty \nu\rangle = \langle T_\infty g,\nu\rangle$ implies $T_\infty g = 0$, thus by assumption $T_\infty g = g$ and therefore $g = 0$.

2) Let J denote the order - ideal generated by $g - T_\infty g$. If $f \in J$ then $-n(g-T_\infty g) \leqq f \leqq n(g-T_\infty g)$ for some $n \in \mathbb{N}$. Thus $-n\, T_\infty(g-T_\infty g) \leqq T_\infty f \leqq n\, T_\infty(g-T_\infty g)$. Since $T_\infty T_\infty g = T_\infty g$ we get $T_\infty f = 0$. In other words: $J \subseteq \ker(T_\infty) = \{x \in E:\ T_\infty x = 0\}$. Since $\ker(T_\infty)$ is closed, $E_{g-T_\infty g} = \bar{J} \subseteq \ker(T_\infty)$.

3) Take $p \in E'_+$ and $f \in E_{g-T_\infty g}$, then $\langle f, T'_\infty p\rangle = \langle T_\infty f, p\rangle = 0$ by (2).
4) Follows from (3) and proposition A1 (4).

The following corollary may help to clarify the relation between the sets C_ε, $C_{p,\varepsilon}$ and $C_{\varepsilon,f}$ defined here, and the corresponding sets C_ε and $C_{x,\varepsilon}$ defined in section 1.

<u>A3. Corollary</u>: Let μ be a finite measure on X and $T: L^\infty(\mu) \to L^\infty(\mu)$ a positive linear Operator, such that the extension of T to some $L^p(\mu)$ $(1 \leq p < \infty)$ is mean ergodic.

1) The following alternative holds:
<u>Either</u> there is a goal function for μ in $L^p(\mu)$,
<u>or</u> there is a strictly positive T'-invariant function in the dual $L^q(\mu)$ $(1/p + 1/q = 1)$.
2) If there is a goal function g for μ in $L^p(\mu)$, then for any 'starting distribution' $p_0 \in L^q(\mu)$, $T'_\infty p_0 \in L^q(\mu)$ is concentrated on $\{x \in X: T_\infty g(x) = g(x)\}$. This set is nonempty, if there is a T - invariant function $0 < h \in L^p(\mu)$.

Proof: We employ theorem A2 for $E = L^p(\mu)$ and $E' = L^q(\mu)$.
1) is immediate from A2 (1).
2) follows from A2 (3):
For any $0 \leq f \in E$, $E^0_f \subseteq E'$ contains exactly those functions in $E' = L^q(\mu)$ which vanish on the measurable set $\{x \in X: f(x) \neq 0\}$.
Now take $f := g - T_\infty g$. The set $\{x \in X: f(x) = 0\}$ is nonempty iff $E^0_f \neq 0$.
To show that this is the case, let $Th = h > 0$ and $\nu \in E'_+$ such that $\langle h, \nu\rangle > 0$, then $\langle h, T'_\infty \nu\rangle = \langle T_\infty h, \nu\rangle = \langle h, \nu\rangle > 0$. Thus $T'_\infty \nu \neq 0$, and by A2 (3) $T'_\infty \nu \in (E_{g-T_\infty g})^0$.

<u>Remark</u>: It would be interesting to find a connection between the sets C_ε defined here (for various functions g) and the conservative (non - dissipative) part of the measure space of a dynamical system (cf. [1], [12]).

<u>Appendix B</u>

Here I want to carry out the construction, scetched in section 2, which allows to express the ergodicity of a system (X, φ) by means of a goal function. For the corresponding thermodynamical terminology see for example [18], especially Part II.
Let $\varphi: X \to X$, and suppose there is a normed φ - invariant measure μ on X. Now we construct a different system (X', φ') by taking

$X' := \{$finite measures on X, absolutely continuous with respect to $\mu\}$, $\varphi'(p)$ defined by $\varphi'(p)(A) := p(\varphi^{-1}(A))$ for every measurable set A and every $p \in X'$.

As a Banach lattice X' is isomorphic to $L^1(\mu)$ (associate with a measure $p \in X'$ it's density f_p with respect to μ). In this sense φ' corresponds to the adjoint T' of the operator $T: f \to f \circ \varphi$: for every measurable $A \subseteq X$ $p(A) = \int 1_A f_p d\mu$ and $\varphi'(p)(A) = p(\varphi^{-1}(A)) = \int 1_{\varphi^{-1}(A)} \cdot f_p d\mu = \int T 1_A \cdot f_p d\mu = = \int 1_A \cdot T' f_p d\mu$. The following proposition will be formulated in terms of the operators T and T', which satisfy $T1 = 1$ and also $T'1 = 1$, since μ is φ-invariant.

The next step is the definition of an appropriate goal function g for $(X', \varphi') \cong (L^1(\mu), T')$. Let $\alpha = \{A_1, \ldots, A_n\}$ be a measurable partition of X. The information gain g_α of the measure ν (or the corresponding density f_ν) with respect to μ for the partition α is defined as follows (cf.2.1): Let $\nu' := \nu / \|\nu\|$ be the normed measure corresponding to ν.

$$g_\alpha : \begin{cases} X' \to \mathbb{R} \\ \nu \to \Sigma_i\ \nu'(A_i) \log(\nu'(A_i)/\mu(A_i)) \end{cases} .$$

For the definitions of mixing and ergodicity in the following proposition see [13],V, Exercise 14 (where ergodicity means irreducibility).

<u>B1.Proposition</u>: Let (X, Σ, μ) be a probability space, $E := L^1(\mu)$, $E' = L^\infty(\mu)$. Let $T: E \to E$ be a linear lattice homomorphism satisfying $T1 = 1$ and $T'1 = 1$. Then T' can be extended continuously from $L^\infty(\mu)$ to $L^1(\mu)$ and the extension will still be called T'. Consider the following statements:

a) T is (strongly) mixing.

b) For every finite measurable partition α of X is g_α a goal function for the system $(L^1(\mu), T')$.

c) T is ergodic.

Then (a) $\Rightarrow$ (b) $\Rightarrow$ (c).

<u>Proof</u>: (a) $\Rightarrow$ (b): If T is strongly mixing, i.e. $T^k \to 1 \otimes 1$, then $\langle T'^k f, 1_A \rangle = \langle f, T^k 1_A \rangle \to \langle f, 1 \rangle \cdot \langle 1, 1_A \rangle = \mu(A)$ for every $A \in \Sigma$ and every $f \in L^1(\mu)$ satisfying $\int f d\mu = 1$. Let α be any partition of X, then also $g_\alpha(T'^k f) \to g_\alpha(\mu) = 0$ (compare 2.3.b). On the other hand $g_\alpha > 0$ can be proved as in 2.3.a. Thus g_α is a goal function.

(b) $\Rightarrow$ (c): Suppose T is not ergodic, then there is an $A \in \Sigma$ such that $\mu(A) \neq 0 \neq \mu(A^c)$ and $T 1_A = 1_A$ and therefore $T 1_{A^c} = T(1 - 1_A) = = T1 - T1_A = 1 - 1_A = 1_{A^c}$. Now take $\alpha = \{A, A^c\}$. Then $g_\alpha = g_\alpha \circ T'$ and g_α cannot be a goal function for $(L^1(\mu), T')$.

Indeed, $T'p(A) = \langle T'p, 1_A \rangle = \langle p, T 1_A \rangle = \langle p, 1_A \rangle = p(A)$ and analogously $T'p(A^c) = p(A^c)$, and therefore $g_\alpha(T'p) = g_\alpha(p)$ for every $p \in L^1(\mu)$.

Proposition B1 remains valid for a positive linear operator T, if the definition of g_α is extended. For a finite set $\beta \subseteq L^\infty(\mu)$ with $\sum_{b\in\beta} b = 1$ and $b \geq 0$ for every $b \in \beta$, define
$$g_\beta : \begin{cases} L^1(\mu) \to \mathbb{R} \\ f \to \sum_{b\in\beta} \langle b, f/\|f\|\rangle \log \dfrac{\langle b, f/\|f\|\rangle}{\langle b,1\rangle} \end{cases}.$$

B2. Proposition: Let (X,Σ,μ) be a probability space, $E:=L^1(\mu)$, $E'=L^\infty(\mu)$. Let $T:E\to E$ be a positive linear operator satisfying $T1=1$ and $T'1=1$. Then T' can be extended continuously from $L^\infty(\mu)$ to $L^1(\mu)$ and the extension will still be called T'. Consider the following statements:

a) T is (strongly) mixing.

b) For every finite set $\beta \subseteq L^\infty(\mu)$ with $b \geq 0$ for every $b \in \beta$ and $\sum_{b\in\beta} b = 1$, $g_\beta(T'^k f) \to 0$ for every $f \in L^1(\mu)$.

c) For every finite set $\beta \subseteq L^\infty(\mu)$ with the above properties, g_β is a goal function for the system $(L^1(\mu), T')$.

d) For every finite measurable partition α of X, $g_\alpha(T'^k f) \to 0$ for every $f \in L^1(\mu)$.

e) T is ergodic.

Then (a) $\Rightarrow$ (b) $\Rightarrow$ (c), (d) $\Rightarrow$ (e).

Proof: (a) $\Rightarrow$ (b): $\langle T'^k f, b\rangle = \langle f, T^k b\rangle \to \langle f,1\rangle\cdot\langle 1,b\rangle = \|f\|\cdot\langle 1,b\rangle$ for every $b \in L^\infty(\mu)$ and every $f \in L^1(\mu)$. Therefore $g_\beta(T'^k f) \to g_\beta(1) = 0$.

(b) $\Rightarrow$ (c): This is clear since $g_\beta > 0$ according to 2.3.a.

(b) $\Rightarrow$ (d): This is trivial, since for every measurable partition α the corresponding set β of characteristic functions $\beta = \{1_A : A \in \alpha\}$ satisfies the properties required in (b).

(c) $\Rightarrow$ (e): Suppose T is not ergodic, then there is a $b \in L^\infty(\mu)$ such that $0 \leq b \leq 1$ and $Tb = b$ and therefore $T(1-b) = T1 - Tb = 1-b$. Now take $\beta = \{b, 1-b\}$, then $g_\beta = g_\beta \circ T'$ and g_β cannot be a goal function for $(L^1(\mu), T')$. Indeed $\langle T'p, b\rangle = \langle p, Tb\rangle = \langle p,b\rangle$ and analogously $\langle T'p, 1-b\rangle = \langle p, 1-b\rangle$ and therefore $g_\beta(T'p) = g_\beta(p)$ for every $p \in L^1(\mu)$.

(d) $\Rightarrow$ (e): Suppose T is not ergodic, then there is a $p \in L^1(\mu)$ such that $T'p = p$, $p \geq 0$, $\|p\| = 1$ and $p \neq 1$. Thus there is a measurable set A, such that $\langle 1_A, p\rangle \neq \langle 1_A, 1\rangle = \mu(A)$ and therefore also $\langle 1_{A^c}, p\rangle \neq \mu(A^c)$. Now take $\alpha = \{A, A^c\}$, then $g_\alpha(p) = g_\alpha(T'p) \neq g_\alpha(\mu) = 0$. Thus $g_\alpha(T'^k p)$ does not converge to 0.

References

[1] Foguel, S.R.: The Ergodic Theory of Markov Processes. New York: Van Nostrand Reinhold 1969.

[2] Glansdorff, P., Prigogine, I.: Physica 30, 351 (1964).

[3] Glansdorff, P., Prigogine, I.: Physica 46, 344 (1970).

[4] Hahn, W.: Stability of Motion. New York: Springer 1967.

[5] Haken, H.: Synergetics. Berlin, Heidelberg, New York: Springer 1977.

[6] Haken, H. (Ed.): Synergetics. Cooperative Phenomena in Multi-Component Systems. Stuttgart: Teubner 1973.

[7] Huang, K.: Statistical Mechanics. New York: J. Wiley 1963.

[8] Marr, D., Poggio, T., Palm, G.: Analysis of a cooperative stereo algorithm. Biol. Cybernetics, in press (1978).

[9] Penrose, O.: Foundations of Statistical Mechanics. London: Pergamon Press 1970.

[10] Pfaffelhuber, E.: Learning and Information Theory, Intern. J. Neuroscience 3, 83 (1972).

[11] Pfaffelhuber, E.: Information theoretic stability and evolution criteria in irreversible thermodynamics. Third European Meeting on Cybernetics and System Research 1976, Vol. III.

[12] Revuz, D.: Markov Chains. Amsterdam: North Holland 1976.

[13] Schaefer, H.H.: Banach Lattices and Positive Operators. Berlin, Heidelberg, New York: Springer 1974.

[14] Schlögl, F.: Zeitschrift für Physik 191, 81 (1966).

[15] Schlögl, F.: Zeitschrift für Physik 243, 303 (1971).

[16] Schlögl, F.: Zeitschrift für Physik 244, 199 (1971).

[17] Sibirsky, K.S.: Introduction to Topological Dynamics. Leyden: Noordhoff 1975.

[18] Stumpf, H., Riekers, A.: Thermodynamik I. Braunschweig: Vieweg 1975, and Riekers, A., Stumpf, H.: Thermodynamik II. Braunschweig: Vieweg 1977.

STRANGE ATTRACTORS IN 3-VARIABLE REACTION SYSTEMS

Otto E. Rössler
Institute for Physical and Theoretical Chemistry
University of Tübingen and
Institute for Theoretical Physics
University of Stuttgart

and

Peter J. Ortoleva
Department of Chemistry
Indiana University, Bloomington, Indiana

Introduction

The Lorenz equation[1] of turbulence generation is a simple 3-variable quadratic differential equation producing complicated 'nonperiodic'[1] behavior. Recent interest in this equation was triggered by a paper of May[2], who showed that similar complicated oscillations can occur in an ecological difference equation. Later, Winfree[3] suggested, following an observation of 'meandering' in a non-stirred chemical reaction system[4], that chemical systems too should be capable of 'chaos'. Subsequently several abstract reaction systems realizing a simpler (non-Lorenzian) type of chaos were described[5]. A complicated 'non-isothermal' abstract reaction system realizing a chemical analogue to the Lorenz equation itself was also found[6].

In the following, two isothermal abstract reaction systems are presented which both produce a type of chaos that implies the presence of a strange attractor over a certain range of parameters. One of the two strange attractors is a Lorenz attractor.

Numerical Results

The following 3-variable nonlinear differential equation is in principle realizable by a chemical reaction network:

$$\begin{aligned} \dot{x} &= a'x + by - cxy - dzx/(x + K_1) \\ \dot{y} &= e + fx^2 - gy - hxy/(y + K_2) \\ \dot{z} &= j + kx - lz. \end{aligned} \qquad (1)$$

The Michaelis-Menten terms on the right-hand side imply that a steady-state approximation of fast-reacting intermediate substances has been assumed to be applicable (cf.[9]). A numerical simulation is provided in Fig.1. This picture is virtually indistinguishable from one obtained earlier with a prototypic non-chemical differential equation[7]. Indeed, Eq.(1) was obtained in a straightforward manner by first applying a linear transformation to this prototypic equation and then providing the Michaelis-Menten type terms. These fractional terms serve the two functions of (a) warranting non-negativity and (b) allowing (by reducing K) an arbitrarily close approximation of the new flow to the prototypic flow.

The following second set of equations has been constructed in a similar way:

$$\begin{aligned} \dot{x} &= ax + by - cxy - (dz + e)x/(x + K_1) \\ \dot{y} &= f + gz - hy - jxy/(y + K_2) \\ \dot{z} &= k + lxz - mz. \end{aligned} \tag{2}$$

A simulation result is shown in Fig.2. This picture again is virtually indistinguishable from one obtained earlier with a corresponding prototypic equation[8].

A Possible Simpler Example

The abstract reaction system depicted in Fig.3 is considerably simpler than those corresponding to Eq.(1) and (2), respectively. The set of rate equations corresponding to the present system is:

$$\begin{aligned} \dot{x} &= a + b'x - (cy + dz)x/(x + K_1) \\ \dot{y} &= ex - fy^2 \\ \dot{z} &= gx - h'z/(z + K_2), \end{aligned} \tag{3}$$

where $b' = b - e - g$ and $h' = hc_o$. A thorough numerical investigation of this system has yet to be performed. Nonetheless, its 'ingredients' are the same as those of Eq.(2)[8]: there is a focus-plus-saddle subsystem in 2 variables (x,y), and a switching-type nonlinearity in the third (z). The latter variable is coupled to the former two in such a way

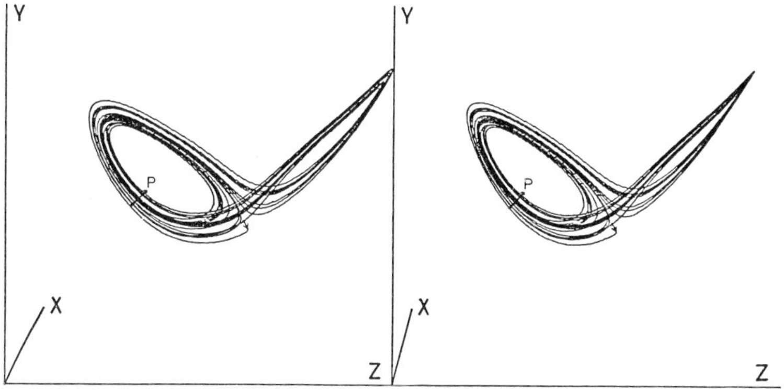

Fig.1 Trajectorial flow of Eq.(1). Numerical simulation using a standard Runge-Kutta-Merson integration routine on a HP 9820A with peripherals. Stereoplots on the basis of parallel projections. (To obtain a stereoscopic picture, fix a pencil in front of the picture at such a distance that, of the four blurred pictures behind it, the two innermost ones merge. Then just wait for them to get in focus.) P = Poincaré cross-section. Parameters: a' = c = d = f = 1, b = 2, e = h = 4, g = 0.1, j = 0.5415, k = 0.06, l = 0.33, $K_1 = K_2$ = 0.002. Initial conditions: x(0) = 1.9997, y(0) = 0.5, z(0) = 2; t_{end} = 46.4. Axes for x: 1...2, for y: 0...2, for z: 1.9...2.1.

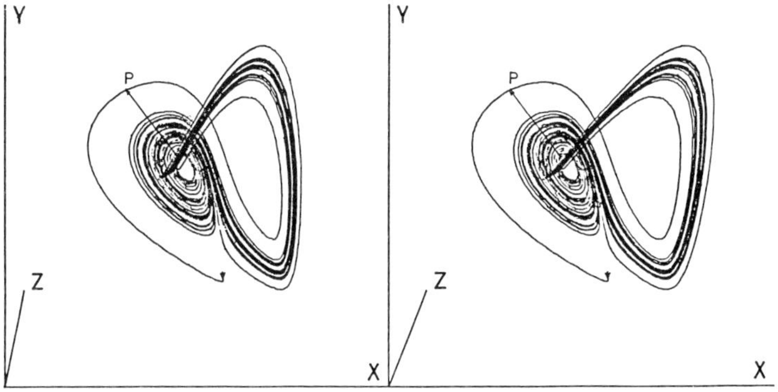

Fig.2 Trajectorial behavior of Eq.(2). Stereoplot as in Fig.1. P = Poincaré cross-section. Parameters: a = 33, b = 150, c = 1, d = 3.5, e = 4815, f = 410, g = 0.6, h = 4, j = k = 2.5, l = 5, m = 750, $K_1 = K_2$ = 0.01. Initial conditions: x(0) = 142.611, y(0) = 15, z(0) = 0.05; t_{end} = 31.7. Axes for x: 0...250, for y: 0...50, for z: 0...750.

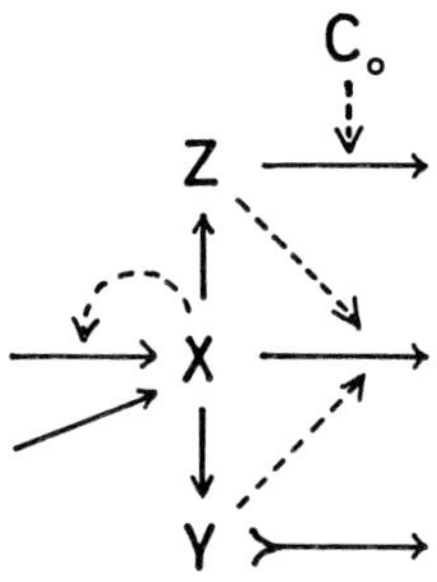

Fig.3 Kinetic scheme of a proposed simpler reaction system allowing strange attractors. Constant pools (sources and sinks) have been omitted as usual. Broken arrows refer to catalysis. For a closely related system, see[10].

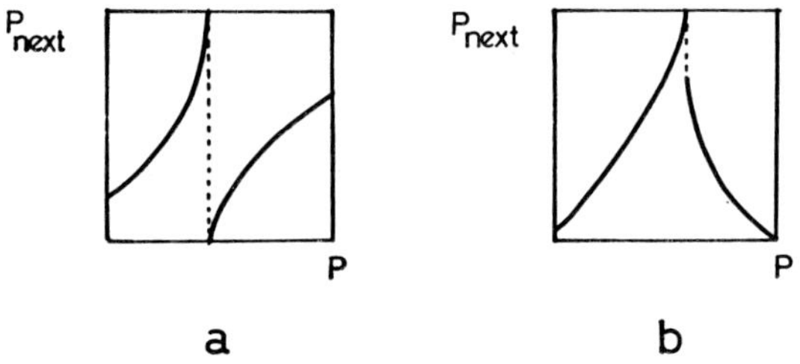

Fig.4 Next re-entry maps (Poincaré cross-sections P) obtained for the trajectorial flows of Figs.1 and 2 when cutting through the left-hand loop of these flows. One-dimensional approximative maps (see[8]). a corresponds to Fig.1 and b to Fig.2.

that the behavior of the whole system is almost the same as that of the subsystem (x,y) as long as x is subcritical with respect to the switching threshold of z; however, following activation of the switch, the trajectories follow a 'roundabout' loop back into the 'non-switched' region[5,10]. The area of 'touch down' (where the 'reinjection'[5] into the non-switched region occurs) can be influenced by manipulating the parameters of Eq.(3). Therefore, there is a finite probability that both the type of flow of Fig.1 and the type of flow of Fig.2 can be obtained from Eq.(3) through parameter optimization.

Discussion

The two types of trajectorial behavior in state space presented above (Fig.1 and Fig.2) can be understood qualitatively by looking at a cross-section through the respective trajectorial flow (most conveniently: through the left-hand loop). If the fact that every such cross-section is in reality two-dimensional is neglected, the one-dimensional approximate maps of Fig.4 are obtained. For a map analogous to the first map (Fig.4a), Guckenheimer[11] recently showed that it implies a strange attractor. Williams[12] extended this result to the (thin) 2-dimensional case and coined the name Lorenz attractor for the family of strange attractors formed. The map of Fig.4b also determines a strange attractor, as can be shown by an analysis analogous to that of Guckenheimer[11] and Williams[12]. Its specific properties have yet to be derived, however (cf.[8]).

Note that no proof has been offered that the flows simulated in Figs.1 and 2 do possess a strange attractor. This would require explicit knowledge of the pertinent cross-sections. Nonetheless, by virtue of the fact that all cross-sections that are 'sufficiently similar' to those of Fig.4 do produce strange attractors, there is no question that Eq.(1) and (2) are both able to produce strange attractors - although perhaps at parameter values slightly different from those chosen in the simulations. The same 'ambiguous' situation incidentally applies to the Lorenz equation itself (see[11]).

All flows governed by a strange attractor are also 'chaotic'[11,8], that is[13], possess an infinite number of unstable periodic solutions plus an uncountable set of nonperiodic ones. The converse is not true, however: the number of 'merely chaotic' systems is much larger, for example in chemistry, than that of systems possessing a strange attractor at the same time[14]. The reason is that most chaotic systems possess, embedded within the 'chaotic attractor'[8] (which is a non-minimal attractor of measure zero[8]), a periodic attractor of mostly large periodicity. This holds true also for the Lorenz equation: only a fraction of its attracting chaotic regimes[15] lack a folding (and hence a 'compression zone'[8] under iteration) and accordingly contain a strange attractor. Physically (and chemically and even numerically) the difference between both sorts of chaotic attractors is irrelevant in most cases, however, because even a very small finite noise level renders the periodic attractor inaccessible. This excessive sensitivity to noise incidentally characterizes not only chaotic attractors, but 'chaotic monoflops'[7] (in which the chaotic regime is not itself an attractor) as well. See [16] and [17] for examples.

As a consequence, the above presented examples of strange attractors in chemistry mainly serve to show that the mathematically 'pure' case can also be realized in 3-variable kinetics.

As a final remark, the fact that more-than-three variable systems are more prone to show strange attractors should be mentioned. In a recent experimental observation of chaotic modes in the Belousov-Zhabotinskii reaction, two-dimensional projections of state space were obtained by plotting the electrochemical potential vs. the potential of a bromide ion sensitive electrode[18] (see also [19]). Among several chaotic pictures obtained, one seems to involve an unstable focus and a saddle point - not too different in this respect from the above pictures.

We thank Art Winfree for discussions.

References

1. E.N. Lorenz, Deterministic Nonperiodic Flow. J. Atmos. Sci. 20,130-141 (1963).

2. R. May, Biological Populations with Nonoverlapping Generations: Stable Points, Limit Cycles, and Chaos. Science 186,645-647 (1974).

3. A.T. Winfree, Personal Communication (1975).

4. A.T. Winfree, Scroll-shaped Waves of Chemical Activity in Three Dimensions. Science 181,937-939 (1973).

5. O.E. Rössler, Chaotic Behavior in Simple Reaction Systems. Z. Naturforsch. 31a, 259-264 (1976).

6. P.J. Ortoleva, Unpublished (1975).

7. O.E. Rössler, Different Types of Chaos in Two Simple Differential Equations. Z. Naturforsch. 31a, 1664-1670 (1976).

8. O.E. Rössler, Continuous Chaos: Four Prototype Equations. In: Bifurcation Theory and Applications (O. Gurel and O.E. Rössler, eds.), Proc. N.Y. Acad. Sci. (in press).

9. F.G. Heineken, H.M. Tsuchiya and R. Aris, On the Mathematical Status of the Pseudo-steady State Hypothesis in Biochemical Kinetics. Math. Biosci. 1, 95-113 (1967).

10. O.E. Rössler, Chaos in Abstract Kinetics: Two Prototypes. Bull. Math. Biol. 39, 275-289 (1977).

11. J. Guckenheimer, A Strange-Strange Attractor. In: The Hopf Bifurcation (J.E. Marsden and M. McCracken, eds.), pp. 368-381. Springer-Verlag: New York 1976.

12. R.F. Williams, The Structure of Lorenz Attractors (preprint).

13. T.Y. Li and J.A. Yorke, Period Three Implies Chaos. Am. Math. Monthly 82, 985-992 (1975).

14. O.E. Rössler, Continuous Chaos. In: Synergetics - A Workshop (H. Haken, ed.), pp. 184-197. Springer-Verlag: New York 1977.

15. O.E. Rössler, Horseshoe-map Chaos in the Lorenz Equation. Phys. Letters 60A, 392-394 (1977).

16. K. Nakamura, Nonlinear Fluctuations Associated with Instabilities in Dissipative Systems. Prog. Theor. Phys. 57, 1874-1885 (1977).

17. J.L. Kaplan and J.A. Yorke, Preturbulence: A Regime Observed in a Fluid Flow Model of Lorenz (preprint).

18. O.E. Rössler and K. Wegmann, Different Types of Chaos in the Belousov-Zhabotinskii Reaction. Thirteenth Symposium on Theoretical Chemistry, Münster, October 1977.

19. O.E. Rössler and K. Wegmann, Chaos in the Zhabotinskii Reaction. Nature (in press).

STRUCTURES OF EXCITATION AND INHIBITION

U. an der Heiden

Lehrstuhl für Biomathematik
Universität Tübingen
Auf der Morgenstelle 28, 7400 Tübingen
Fed. Rep. Germany

Summary. Neural networks are modelled in the form of systems of integral equations. These equations are derived from general principles of neural activity and should be flexible enough for data fitting. Several well known models appear as specializations.

Introduction. The generation, transport and processing of information within the nervous system is achieved by the parallel activity of a large number of interacting neurons. In the following a rather flexible model for the interaction of neurons and for the activities in neural networks is developed. The model is derived from basic principles and properties common to nearly all types of neurons so far investigated (section 1). A characteristic feature of the theory presented here is its formal generality. It does not describe individual neurons or individual networks but gives a framework, i.e. the general form of a neural network. Special networks can be obtained by specializing the functions occuring in the model through experimental measurements or through derivation from other more special models of neurons or interaction principles.

It turns out that many of the classical and recent models are special cases of this system. In this way a precise relation between all those models is established which allows continuous transition from one to the other (see section 2). However, the main point is the openness of the theory to the incorporation of experimental data. In my opinion many of the models in the literature are too rigid so that the desirable congruence between theory and experiments is excluded from the beginning.

The theory is given in the form of integral equations. In order to show that they are without contradiction existence - and uniqueness theorems are proved in section 3.

1. Neuronal Structures

a) The Conversion of Impulse Frequencies into Generator Potentials

In general two neurons A and B are not connected by a single synapse. On the contrary there are examples where A makes contact with B via a number of synapses of the order 10^2. Therefore in general it is not practicable to take into account monosynaptic transfer characteristics. What seems to be within the reach of experimental registration and mathematical description is the total influence exerted from A on B. This influence is determined among other things by the geometry of the dendritic and axonal arborizations and the localization of the synapses. We refer in this context to the work of W. Rall.

Let us assume that a single impulse (= spike) is generated at the axon hillock of neuron A at time $t = 0$, i.e. the impulse frequency (= spike frequency = number of spikes per second) of A equals $\delta(t)\ \mathrm{sec}^{-1}$ for $-\infty < t < \infty$, where $\delta(t)$ denotes Dirac's function. Let for $t < 0$ the membrane potential of B be equal to the resting potential v_{B0}. Then the response of the membrane potential v_B near to the axon hillock of B (this is called the generator potential or slow potential of B) to the impulse in A is described by a function

$$v_B(t) = v_{B0} + h_{BA}(t) \tag{1}$$

with the impulse response function h_{BA} satisfying $h_{BA}(t) = 0$ for $t < 0$. If $h_{BA}(t) \geq 0$ we have excitation and if $h_{BA}(t) \leq 0$ we have inhibition of B by A. However, we mention that there may be impulse responses with alternating sign (see [11]).

Under the assumption that the conversion of the impulse frequency x_A of cell A into the generator potential v_B of cell B is that of a linear time-invariant, causal system it follows from elementary principles of system theory that v_B can be computed from x_A by forming the convolution of x_A with h_{BA}:

$$v_B(t) = v_{B0} + \int_{-\infty}^{t} x_A(t')\, h_{BA}(t-t')dt'. \tag{2}$$

With the abbreviation

$$f * h = \int_{-\infty}^{t} f(t')\, h(t-t')dt'$$

equation (2) reads

$$v_B = v_{B0} + x_A * h_{BA}. \tag{3}$$

The assumption of linearity in general gives only a first order approximation of the real system. Due to thresholds for the activation of synapses and exhaustion of transmitter substance for large values of x_A the influence of x_A on v_B obeys a nonlinearity $S_{BA}(x_A)$ with $S_{BA} : \mathbb{R}_+ \to \mathbb{R}$ being in most cases an increasing and bounded function of x_A. The function S_{BA} incorporates properties of the synaptic transmission from A to B, and equation (3) has to be refined to

$$v_B = v_{Bo} + S_{BA}(x_A) * h_{BA}, \tag{4}$$

with $S_{BA}(x_A)(t) = S_{BA}(x_A(t))$.
The function h_{BA} can be interpreted in this context as a temporal weighting factor indicating the strength of the influence of the past values $x_A(t')$, $t' \leq t$, on the present value $v_B(t)$.

Until now we did not specify whether the neuron A is different from the neuron B. Indeed generally after the generation of an impulse in B there is a decrease (self-inhibition) of the generator potential in B. We assume that the effect of x_B on v_B can be separated at least theoretically from the effect of x_A on v_B if A and B are different and that the total effect obeys the principle of superposition. Then the temporal development of v_B in a system of two neurons A and B is described by

$$v_B = v_{Bo} + S_{BB}(x_B) * h_{BB} + S_{BA}(x_A) * h_{BA}. \tag{5}$$

The superposition principle will be relativated in part b) of this section.

b) <u>The Conversion of Generator Potentials into Impulse Frequencies</u>

In spike generating neurons the generator potential is transformed at the axon hillock into a series of impulses which will travel down the axon of the neuron.
A spike is generated if and only if the generator potential $v(t)$ exceeds a threshold value $\theta = \theta(t)$, which is a function of time. The variation of θ with time can be derived theoretically from the models of Hodgkin-Huxley [7] and Fitzhugh [4] (see there p. 449) for the generation of the nervous impulse, however only as far as purely qualitative characteristics are concerned. Quantitative realistic values can only be obtained by measurements.
The qualitative behavior of θ is in general as follows.

Let us assume an impulse is generated at time $t_o = 0$. Then during the absolute refractory period $(0, r)$ the neuron is unable to produce a new spike, i.e. $\theta(t) = \infty$ for $0 < t < r$.

Afterwards the threshold decays continuously to an asymptotic value θ_o, the resting threshold, provided that no new impulse is triggered, i.e. that $v(t) < \theta(t)$ for $t \geq t_o + r$.

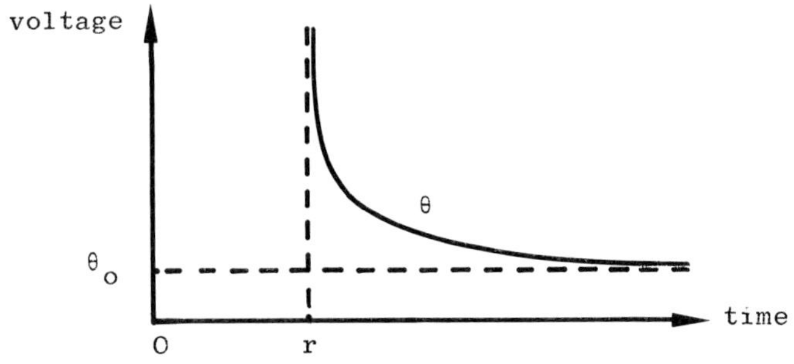

The time during which the threshold is finite and significantly greater than θ_o is called the relative refractory period of the neuron. The first spike after that at t_o occurs at the first time $t = t_1 > t_o$ satisfying $v(t) \geq \theta(t)$.

After t_1 the threshold develops just as after t_o. A neuron will show spontaneous firing if its resting potential v_o is below the resting threshold θ_o, i.e. if $v_o < \theta_o$.

Since we are not interested in the details of a single spike but in the spike frequency, i.e. the number of spikes per second, the variation of v within a time interval which corresponds to the duration of a single spike (about 1 msec) cannot be taken into account.

Therefore in the following we make the assumption that v does not vary too rapidly within 1 msec (otherwise one has to pass over to mean values of v with respect to 1 msec).
Then the following functional relation $\bar{S}$ between the generator potential v and the impulse density x at the beginning of the axon of a cell is a consequence of the temporal development of the threshold:

$$x(v) = \bar{S}(v) = \begin{cases} 0 & \text{if} \quad v \leq \theta_o \\ 1/\theta^{-1}(v) & \text{if} \quad v > \theta_o . \end{cases} \tag{6}$$

Here θ^{-1} denotes the inverse function of θ, i.e. $\theta^{-1}(v) = t$ if $\theta(t) = v$, $v > \theta_o$.

The voltage-to-frequency conversion function $\bar{S}$ is different for dif-

ferent types of neurons, sometimes also differences between neurons of the same type are significant.
It follows from (6) and the properties of θ that $\bar{S} = \bar{S} : \mathbb{R} \to [o, r^{-1}]$ is a nonnegative, bounded, nondecreasing function. If θ is continuous on the interval (r, ∞), then $\bar{S}$ is continuous on the whole real line $\mathbb{R}$. According to the relations (5) and (6) the impulse frequency $x_B(t)$ in a neuron B coupled to a neuron A with frequency $x_A(t)$ is

$$x_B = \bar{S}_B(v_{Bo} + S_{BB}(x_B) * h_{BB} + S_{BA}(x_A) * h_{BA}), \qquad (7)$$

where $\bar{S}_B$ denotes the voltage-to-frequency conversion function of the cell B.

On the other hand, since $x_A = \bar{S}_A(v_A)$ and $x_B = \bar{S}_B(v_B)$, the potential v_B of cell B is related to the potential v_A of cell A by the integral equation

$$v_B = v_{Bo} + [S_{BB} \circ \bar{S}_B(v_B)] * h_{BB} + [S_{BA} \circ \bar{S}_A(v_A)] * h_{BA}, \qquad (8)$$

($f \circ g(t)$ means $f(g(t))$).
The equations (7) and (8) are two equivalent descriptions of the influence of the activity in one cell on the activity in another cell. With

$$T_{BA} = S_{BA} \circ \bar{S}_A \qquad (9)$$

equation (8) is simplified to

$$v_B = v_{Bo} + T_{BB}(v_B) * h_{BB} + T_{BA}(v_A) * h_{BA}. \qquad (10)$$

c) Structures of Neural Networks

We assume that there is given a network of a finite number of neurons $A_1, A_2, \dots, A_n$, all obeying the principles discussed in a) and b) of this section. It follows from equations (7) and (10) that the activity in this network is determined by the system of n integral equations

$$x_i(t) = \bar{S}_i(v_{io} + \sum_{j=1}^{n} S_{ij}(x_j) * h_{ij} + E_i * \bar{h}_i), i = 1,2,\dots,n, \qquad (11)$$

or equivalently in terms of the slow potentials

$$v_i(t) = v_{io} + \sum_{j=1}^{n} T_{ij}(v_j) * h_{ij} + E_i * \bar{h}_i, \; i = 1,2,\dots,n. \qquad (12)$$

The subscript i assigns the corresponding variable to the neuron A_i, the pair ij labels the terms characterizing the influence of the j-th neuron on the i-th neuron. The quantity $E_i = E_i(t)$ represents the external input to the neuron A_i, and $\bar{h}_i$ describes the response of the neuron to an external impulse shaped input. Eventually a nonlinearity has also to be implemented into the external input behavior in an obvious way.

In the systems (11) and (12) the individual neurons are conceived as discrete units. However under several circumstances it is more appropriate to consider the neurons as members of a continuum. Such circumstances may be

(α) the number n is so large and the different neurons are so similar that it is mathematically justified to pass over to a continuous model,

(β) the neurons under consideration form an extremely homogenuous, densily packed and relatively isolated subsystem of the nervous system (e.g. if the behavior of a single neuron is not very significant).

We do not want to give precise conditions for the realization of (α) and (β). It is a matter of fact that there are different types of neurons in the brain, the neurons within one type having very similar properties. Moreover the neurons of the same type are often distributed in the nervous tissue in a very regular order and also the interconnections are formed very regularly (examples are the retina of the eye or the visual cortex with its different layers). In this situation a continuous model is appropriate.

Along these lines we assume that we have m different types of neurons. We label the different types by the index k, k = 1, 2, ..., m. The neurons of type k are distributed in a region R_k of the brain. Geometrically R_k is interpreted as a 1-, 2-, or 3- dimensional manifold according as the neurons are ordered in a chain, a layer or a cluster. The geometrical relations of neighborhood (the topolgy of R_k) are given by the arrangement of the neurons in the tissue.
A neuron of type k is identified as point s_k in the domain R_k, $s_k \in R_k$. The impulse density and the membrane potential at s_k at time t is denoted by $x(s_k, t)$ and $v(s_k, t)$ resp.
The whole neural network including the interactions of the neurons within one type and between different types is described by the following set of equations as a consequence of the equation (7):

$$x(s_k,t) = \bar{S}(s_k, [v_o(s_k) + E(s_k,.) * \bar{h}(s_k,.) + \sum_{l=1}^{m} \int_{R_l} S(s_k,s_l',x(s_l',.)) * h(s_k,s_l',.)ds_l' + \tilde{S}(s_k,x(s_k,.))*\tilde{h}(s_k,.)]), \tag{13}$$

where $s_k \in R_k$, $k = 1, 2, \ldots, m$, (~ refers to self-inhibition).

Equivalent to the description (13) there exists a representation with respect to the potentials which follows from equation (10):

$$v(s_k,t) = v_o(s_k) + E(s_k,.) * \bar{h}(s_k,.) + \tilde{T}(s_k,v(s_k,.)) * \tilde{h}(s_k,.) + \sum_{l=1}^{m} \int_{R_l} T(s_k,s_l',v(s_l',.) * h(s_k,s_l',.) \, ds_l'. \tag{14}$$

The functions S and T are related by the equations

$$T(s_k,s_l',v(s_l',t)) = S(s_k,s_l',\bar{S}(s_l',v(s_l',t)) = S(s_k,s_l',x(s_l',t)), \tag{15}$$

and

$$\tilde{T}(s_k,v(s_k,t)) = \tilde{S}(s_k,\bar{S}(s_k,v(s_k,t)) = \tilde{S}(s_k,x(s_k,t)), \tag{16}$$

with $s_k \in R_k$ and $s_l' \in R_l$.

In the following section it will be shown that the systems (12) and (14) include several well known models for neural networks as specializations. In this way a relation between these various models is exhibited which allows a continuous transition from one model to the other. The last section contains existence- and uniqueness-theorems for the integral equations (11), (12), (13) and (14).

2. On the Relation between Several Models for Neural Networks

a) The Hartline-Ratliff Equations

In principle the functions S and h should be measurable experimentally. This has been done for the retinal network of the complex eye of Limulus (horseshoe crab) by the Hartline-Ratliff school [11]. The interesting variables are the potentials $v_i(t)$ and the impulse frequencies $x_i(t)$ in the eccentric cells of the single eyes (= ommatidia). If these cells are numbered from 1 to n then we obtain a finite network of the form (11). It turns out [10] that the measurements are well approximated by the expressions

$$\bar{S}_i(v) = b_i \max(0, v - \theta_i),$$

$$S_{ij}(x) = \max(0, x - r_{ij}),$$

$$h_{ij}(t) = H(t-\tau) \,.\, K_{ij}\delta_1 \, e^{-(t-\tau)/\delta_1} \quad \text{for } i \neq j \tag{17}$$

$$h_{ii}(t) = H(t) \,.\, K_{ii}\delta_2 \, e^{-t/\delta_2},$$

where θ_i and r_{ij} are constant threshold values, H = Heaviside-function ($H(t) = 0$ für $t < 0$, $H(t) = 1$ for $t \geq 0$), $\tau \approx 0.1$ sec is a constant delay in the impulse response, δ_1, δ_2 time constants. The interaction constants K_{ij} are negative: the Limulus retina represents a network with lateral inhibition.

The system [(11), (17)] has been investigated by Coleman & Renninger [1], [2] mathematically under certain homogeneity conditions. The steady state solutions $\bar{x}_i$ = const. of [(11), (17)] are just the well known Hartline-Ratliff equations

$$\bar{x}_i = \max(0, e_i - \sum_j K_{ij} \max(0, \bar{x}_j - r_{ij})), \quad i = 1, \ldots, n. \tag{18}$$

Equations (18) are also a special case of (11) with $\bar{S}_i$ and S_{ij} defined as in (17) and with $h_{ij} = K_{ij}\,\delta(t)$, where $\delta(t)$ again denotes Dirac's function.

b) A Model of J.D. Cowan

We now assume the functions h in the equations (12) have the form

$$h_{ij}(t) = A_{ij}\, e^{-t/\mu}\, H(t), \quad \bar{h}_i = A_i\, e^{-t/\mu}\, H(t) \tag{19}$$

with a universal neural time-constant μ and constants A_{ij}. Moreover we assume $S_{ij}(x_j) = x_j$ and $\bar{S}_i$ to be sigmoid functions. Then according to (9) and (12) the membrane potentials v_i satisfy

$$v_i(t) = v_{io} + \int_{-\infty}^{t} \left(\sum_{j=1}^{n} A_{ij}\, \bar{x}_j(t') + A_i E_i(t')\right) e^{-(t-t')/\mu}\, dt'. \tag{20}$$

Derivation of this expression with respect to t results in

$$\frac{dv_i}{dt} = -\frac{1}{\mu} \int_{-\infty}^{t} \left(\sum_{j=1}^{n} A_{ij}\, x_j(t') + A_i E_i(t')\right) e^{-(t-t')/\mu}\, dt' +$$

$$+ \sum_{j=1}^{n} A_{ij}\, x_j(t) + A_i E_i(t),$$

i.e.

$$\mu \frac{dv_i}{dt} = -v_i(t) + e_i(t) + \sum_{j=1}^{n} A_{ij}\, \bar{S}_j(v_j), \quad i = 1,2,\ldots,n \tag{21}$$

with $e_i = v_{io} + \mu\, A_i E_i$.

In an entirely similar manner field equations are obtainable from equations (14) instead of (12): For $s_k \in R_k$, $k = 1, \ldots, m$:

$$\mu \frac{\partial v(s_k,t)}{\partial t} = -v(s_k,t) + e_k(s_k,t) + \sum_{l=1}^{m} \int_{R_l} A(s_k,s_l')\, \bar{S}(s_l',v(s_l',t))ds_l. \tag{22}$$

Systems (21) and (22) represent the model derived by Cowan and Feldman (s. [3] p. 37) which they applied to motoneuron pools and to the brain stem respiratory center. Closely related to this model are those considered by Griffith [5], Morishita, Yajima, Tokura [9], [13] and Stein et al. [12]. They are all obtainable by specializations of the equations (11) - (14). We do not go into details.

c) <u>The Model of McCulloch and Pitts</u>

In contrast to the models of part a) and b) of this section in the model of McCulloch and Pitts [8] time is discretized into entire multiples $t = 0, \tau, 2\tau, 3\tau, \ldots$ of a unit time τ which is in the order of 1 msec. At every such time t a neuron is either in the state 1 ("firing") or 0 ("resting"). In a finite network of such "logical neurons" the state $x_i(t)$ of neuron i, $i = 1, \ldots, n$, at time t is determined by the state of the network at time $t - \tau$ and the external input $E_i(t-\tau)$:

$$x_i(t) = H(E_i(t-\tau) + \sum_{j=1}^{n} K_{ij}\, x_j(t-\tau) - \theta_i), \tag{23}$$

where the K_{ij} are interaction constants and θ_i constant threshold values. Equations (23) are subsumable under the system (11) by specializing

$$\bar{S}_i(v) = H(v-\theta_i),\; S_{ij}(x_j) = x_j,\; h_{ij}(t) = a_{ij}\, \delta(t-\tau), \bar{h}_i = \delta(t-\tau), \tag{24}$$

where H and δ denote the Heaviside- and Dirac-function respectively. It does not matter that system (11) satisfying (24) is defined for all t, since the state at time t depends only on the state at time $t-\tau$.

It is possible to give within the framework of system (11) a continuous

transition of the McCulloch-Pitts model into the Hartline-Ratliff model (characterized by the equations (17)): Only for simplicity we assume in the following that $K_{ii} = 0$ and $r_{ij} = 0$.
Define for each $\lambda > 0$ a specialization of system (11) by

$$\bar{S}_i^{(\lambda)}(v) = \begin{cases} \lambda\, b_i \max(0,\, v-\theta_i) & \text{for } v < \theta_i + 1/(\lambda b_i) \\ 1+b_i\,(v-\theta_i-1/(\lambda b_i))/\lambda & \text{for } v \geq \theta_i + 1/(\lambda b_i), \end{cases} \tag{25}$$

$$h_{ij}^{(\lambda)}(t) = H(t-\tau)\, K_{ij}\, \delta_1 \lambda\, e^{-(t-\tau)/(\delta_1 \lambda)}, \quad S_{ij}(x_j) = x_j.$$

Then for $\lambda = 1$ we have system (11) with conditions (17), the Hartline-Ratliff model, and for $\lambda \to \infty$ we obtain system (11) with conditions (24), the McCulloch-Pitts model.

3. Existence- and Uniqueness Theorems

This section is concerned with conditions sufficient for the existence and uniqueness of solutions of the systems (11) and (13). The conditions to be given are not the most general but they should cover those cases which are likely to occur in the applications.

We derive the results primarily for the system (13), the arguments and conditions being very similar for system (11).

An <u>initial condition</u> for system (13) is a system of continuous and bounded functions $x : R_k \times (-\infty, 0) \to \mathbb{R}$, $k = 1, 2, \ldots, m$. A solution of system (13) to such an initial condition is a system of functions $x(s_k, t)$, $s_k \in R_k$, $t \geq 0$, satisfying the equations (13) for all t with $0 \leq t < \infty$.

<u>Theorem</u>. The system described by the equations (13) has for every initial condition a uniquely determined continuous solution if the following five conditions hold for $k = 1, 2, \ldots, m$:

(i) R_k is a closed and bounded subset of some finite-dimensional Euklidean space $\mathbb{R}^n$,

(ii) $E(s_k,t) = 0$ for $t < 0$, $E(s_k,t)$ is continuous on $R_k \times [0, \infty)$,

(iii) $v_0(s_k)$ is continuous on R_k,

(iv) the functions $\bar{S}(s_k,w)$, $S(s_k,s_1',w)$ and $\tilde{S}(s_k,w)$ are continuous on $R_k \times \mathbb{R}$, $R_k \times R_1 \times \mathbb{R}$ and $R_k \times \mathbb{R}$ respectively and satisfy a uniform Lipschitz condition with respect to the variable w.

(v) the functions $h(s_k,s_1',t)$, $\bar{h}(s_k,t)$, $\tilde{h}(s_k,t)$ are continuous and uniformly bounded on $R_k \times R_1 \times [0, \infty)$, $R_k \times [0, \infty)$, $R_k \times [0, \infty)$

respectively, moreover the integrals

$$\int_0^\infty |h(s_k,s_l',t)|\,dt,\ \int_0^\infty |\bar{h}(s_k,t)|\,dt \text{ and } \int_0^\infty |\tilde{h}(s_k,t)|\,dt \text{ are finite.}$$

Remark. A function $f(u,w)$ satisfies a uniform Lipschitz condition with respect to w means that there is a constant $L > 0$ such that

$$|f(u,w) - f(u,\tilde{w})| < L\,|w-\tilde{w}|$$

for all (u,w), $(u,\tilde{w})$ in the domain of f.

Proof of the theorem. Let $|R_k|$ be the content of R_k, $K > 0$ a bound for the functions in condition (v), i.e.

$|h(s_k,s_l',t)| \le K$, $|\bar{h}(s_k,t)| \le K$ and $|\tilde{h}(s_k,t)| \le K$, and let ε be a number with

$$0 < \varepsilon < (L^2K(1 + \sum_{k=1}^{m} |R_k|))^{-1}.$$

Let $t_0 \ge 0$ and assume that there exist uniquely determined functions $x(s_k,t)$ continuous on $R_k \times [0,t_0]$, coinciding on $R_k \times (-\infty,0)$ with the initial condition, and satisfying (13) on $[0,t_0]$. The theorem will be proved if $x(s_k,.)$ can be extended uniquely to the interval $I = [t_0,t_0+\varepsilon]$ such that $x(s_k,t)$ is continuous on $R_k \times I$ and satisfies (13). To prove the last statement let R be the topological sum of $R_1, R_2, \dots, R_m$. Then $B(R) = \{f : R \to \mathbb{R} : f \text{ is continuous}\}$ is a Banachspace with respect to the norm $\|f\| = \max\{|f(s)| : s \in R\}$. The set $M = \{F : I \to B(R) : F \text{ is continuous}\}$ defines a Banachspace with norm $\|F\| = \max\{\|F(t)\| : t \in I\}$.

An operator $T : M \to M$ is determined in the following way. For $F \in M$ let $y(s_k,t) = x(s_k,t)$ if $t < t_0$ and $y(s_k,t) = F(t)\,(s_k)$ if $t \in I$. For $t \in I$ let $TF(t)\,(s_k)$ be equal to the expression obtained from the right hand side of equation (13) by replacing $x(s_l',.)$ and $x(s_k,.)$ by $y(s_l',.)$ and $y(s_k,.)$ respectively. The continuity and boundedness assumptions in (i) - (v) and on the initial condition imply that TF is defined and that $TF \in M$.

For any two functions F, $G \in M$ we have

$$|T\,F(t)\,(s_k) - T\,G(t)\,(s_k)| \le$$

$$\le L\,(\sum_{l=1}^{m} \int_{R_l} \int_{t_0}^{t} L|F(t')(s_l') - G(t')(s_l')| \cdot |h(s_k,s_l',t-t')|\,dt'ds_l' +$$

$$+ \int_{t_o}^{t} L\,|F(t')(s_k) - G(t')(s_k)| \cdot |\tilde{h}(s_k, t-t')|\, dt')$$

$$\le L^2 K(t-t_o)\ (\sum_{l=1}^{m} |R_l| + 1)\ \|F-G\| \le L^2 K\ (\sum_{l=1}^{m} |R_l| + 1)\ \|F-G\|.$$

Since these inequalities hold for all $s_k \in R_k$ and for all $t \in I$ and because of the choice of ϵ, the inequality

$$\|TF - TG\| \le \varkappa \|F - G\|$$

holds with a constant $\varkappa < 1$. Therefore T is a contraction mapping and Banach's fixed point principle implies that there is a unique function $F_o \in M$ with $TF_o = F_o$. Obviously $x(s_k, t) = F_o(t)(s_k)$ is the unique extension of $x(s_k, t)$ onto the interval $[t_o, t_o+\epsilon]$. Q.E.D.

For the finite-dimensional model (11) conditions on the functions S_i, S_{ij}, h_{ij}, h_i and E_i analogous to those given in (iv), (v) and (ii) of the previous theorem ensure the existence and uniqueness of solutions of system (11). Another question is whether (11) or (13) have stationary solutions.

Let us consider this problem in more detail for the system (11). A stationary solution is a solution which does not depend on time, i.e. $x_i(t) = \bar{x}_i$ = constans for $i = 1, 2, \ldots, n$.

Stationary solutions can in general only be expected if there is no input to the network or if the input is itself stationary. In the following we assume $E_i(t) = \bar{E}_i$ = constans, $i = 1, 2, \ldots, n$, moreover it is necessary to assume

$$H_{ij} = \int_0^\infty h_{ij}(t)\, dt < \infty \text{ and } \bar{H}_i = \int_0^\infty \bar{h}_i(t)\, dt < \infty,\ i,j = 1, 2, \ldots, n.$$

The system (11) has a stationary solution $\bar{x}_1, \bar{x}_2, \ldots, \bar{x}_n$ if and only if

$$\bar{x}_i = \bar{S}_i(e_i + \sum_{j=1}^{n} H_{ij}\, S_{ij}(\bar{x}_j)),\ i = 1, 2, \ldots, n, \tag{26}$$

where $e_i = v_{io} + \bar{E}_i \bar{H}_i$.

<u>Theorem</u>. Let e_i and H_{ij} be arbitrary constants and let $\bar{S}_i : \mathbb{R} \to [0,\infty)$, $S_{ij} : [0,\infty) \to \mathbb{R}$ be continuous functions for $i,j = 1, 2, \ldots, n$. Then the system (26) has at least one nonnegative solution, if the func-

tions $\bar{S}_i, i = 1, 2, \ldots, n$ are bounded.

Proof. We apply Brouwer's fixed point theorem: If $\Omega \subset \mathbb{R}^n$ is a bounded, closed, convex set and $F : \Omega \to \Omega$ is a continuous function then there is a fixed point $\bar{x}$, $F\bar{x} = \bar{x}$. Let

$$M_i = \sup \{ \bar{S}_i(x) : x \in \mathbb{R} \} \text{ and define}$$

$$\Omega = \{ (x_1, x_2, \ldots, x_n) \in \mathbb{R}^n : 0 \leq x_i \leq M_i, \ i = 1, 2, \ldots, n \}.$$

Let $F = (F_1, F_2, \ldots, F_n)$ be defined by

$$F_i (x_1, x_2, \ldots, x_n) = \bar{S}_i \left(e_i + \sum_{j=1}^{n} H_{ij} S_{ij} (x_j)\right).$$ According to

Brouwer's theorem there is a fixed point $\bar{x} = (\bar{x}_1, \ldots, \bar{x}_n)$ of F. $\bar{x}$ is a solution of (26). Since $\Omega \subset \mathbb{R}^n_+$ we have $\bar{x}_i \geq 0$, Q.E.D.

The stationary solutions are in general not uniquely determined. A counter example is given in [6] for the Hartline-Ratliff equations. Statements similar to the previous theorem can be made for the infinite-dimensional network (13) by applying more general fixed point theorems.

References

1. Coleman, B.D., Renninger, G.H.: Periodic solutions of a nonlinear functional equation describing neural interactions. Istituto Lombardo (Rend Sc.) A 109, 91-111 (1975)
2. Coleman, B.D., Renninger, G.H.: Theory of the response of the Limulus retina to periodic excitations. J. Math. Biol. 3, 103-120 (1976)
3. Feldman, J.L., Cowan, J.D.: Large scale activity in neural nets. Biol. Cyb. 17, 29-51 (1975)
4. Fitzhugh, R.: Impulses and physiological states in theoretical models of nerve membrane. Biophys. J. 1, 445-466 (1961)
5. Griffith, J.S.: A field theory of neural nets I, II. Bull. Math. Biophys. 25, 111-120 (1963) & 27, 187-195 (1965)
6. Hadeler, K.P.: On the theory of lateral inhibition. Biol Cyb. 14, 161-165 (1974)
7. Hodgkin, A.L., Huxley, A.F.: A quantitative description of membrane current and its application to conduction and excitation in nerve. J. Physiol. (London) 117, 500-544 (1952)
8. McCulloch, W.S., Pitts, W.H.: A logical calculus of ideas immanent in nervous acitivity. Bull. math. Biophys. 5, 115-133 (1943)
9. Morishita, I., Yajima, A.: Analysis and simulation of networks of mutually inhibiting neurons. Biol. Cyb. 11, 154-165 (1972)

10. Ratliff, F., Knight, B.W., Graham, N.: On tuning and amplification by lateral inhibition. Proc. Nat. Acad. Sci., 62, 733-740 (1969)

11. Ratliff, F. (ed.): Studies on excitation and inhibition in the retina. Chapman and Hall, London (1974)

12. Stein, R.B., Leung, K.V., Mangeron, D., Oguztöreli, M.N.: Improved neuronal models for studying neural networks. Biol. Cyb. 15, 1-9 (1974)

13. Tokura, T., Morishita, I.: Analysis and simulation of double-layer neural networks with mutually inhibiting interconnections. Biol. Cyb. 25, 83-92 (1977)

A NEW APPROACH TO SYNAPTIC INTERACTIONS

T. Poggio and V. Torre*

0. Introduction

Two types of signals are used by nerve cells: graded potentials and action potentials. The graded potentials are passively conducted and play an important role at special regions such as at junctions between cells. The action potentials are conducted, regenerative impulses that do not decrease over distances. Recent research suggests that localized graded potentials may play an essential information processing role and underly the integrative function of neurons; action potentials mainly serve a transmission and transduction purpose (Kuffler and Nicholls, 1977).

A quantitative approach to passive neural integration has been developed during the past 20 years, especially by W. Rall and his coworkers. This approach deals with the mechanisms by which charge crosses synaptic junctions in dendrites and then spreads to the soma where it combines with charge spreading from other dendrites. Cable models of neurons are essential tools in evaluating the role of dendritic geometry, dendritic membrane properties and synaptic location on dendrites for the accumulation and spread of charge in neurons. Although a large body of theoretical information is already available, current approaches are essentially restricted to the linear case in which synaptic currents are considered to be the input signal to a cable-like neuron. In this case the postsynaptic voltage depends linearly on the excitatory or inhibitory synaptic

---

* Permanent address: Università di Genova, Istituto di Scienze Fisiche, Genova, Italia

inputs and their combination. In reality, however, postsynaptic effects consist primarily of localized conductance changes for specific ions (see the equivalent circuit of fig. 1). In general, the resulting synaptic currents, which depend on the particular ions, their reversal potential and the instantaneous value of the membrane potential, are not proportional to the input conductance changes. The transduction from conductance changes to somatic potential is a nonlinear process, and its theoretical treatment is restricted to numerical studies (see for instance Rall, 1964; Rinzel and Rall, 1974; Jack, Noble and Tsien, 1975). This is in sharp contrast with the linear case where a general approach is available, in terms of the Green function (and Laplace transform) of the appropriate linear cable equation (see Rinzel and Rall, 1974; Jack, Noble and Tsien, 1975; and especially Butz and Cowan, 1974).

In this paper we outline a method to describe quantitatively nonlinear effects of synaptic conductance changes. Our approach reduces the nonlinear case to the linear one. Active membrane processes cannot be dealt with by our theory in its present form.

Beside its contribution to the cable theory of neurons our approach may also have other implications. In this framework the information processing role of graded potentials and synapses may become transparent. We conjecture that synapses rather than spike encoding neurons are the basic computing elements in the nervous system. This idea is at odds with the traditional view, in which "logical" threshold neurons are the only computational elements in nervous systems and synapses perform a linear, arithmetic sum of positive and negative input components.

This paper provides some theoretical tools for connecting specific computations with the geometrical arrangements, membrane properties and location of synapses (see also Blomfield, 1974). In the framework of our present approach we have recently proposed (Torre and Poggio, 1977; see also section 7) a specific synaptic mechanism involving synapses of the diade type as the basic neural "hardware" that underlies movement detection. If this hypothesis turns out to be correct - experimental evidence should be soon available, for instance in the vertebrate retina - it would then be tempting to regard our approach as providing some useful tools for the understanding of nervous information processing elements and circuits.

The plan of the paper is as follows. Section 1 introduces the basic equation describing the (nonlinear) transduction conductance changes - local potential. The solution of the equation is given in section 2 and its convergence is discussed in section 3. The practical implications of the approach and the underlying biophysical hypotheses are formulated in section 4. Specific examples are discussed throughout. Some formal aspects are briefly outlined in section 5. The use of Feynman-like graphs is also proposed (section 6). The link with information processing is suggested in section 7. In particular, elementary computations are connected (in an indicative way) to possible synaptic morphologies.

1. Basic Equation

Let us consider a neuron with its dendritic tree in the framework of passive cable theory (see Jack, Noble and Tsien, 1975). We assume that N synapses control corresponding ionic channels with batteries $E_1,\dots,E_N$ through conductance changes $g_1,\dots,g_N$ at locations $1,\dots,N$. Two or more of these locations may in fact coincide for all practical purposes. $K_{ij}(\tau)$ denotes the impulse response relating the potential change at location i to a delta pulse of current at location j. $K_{ij}(\tau)$ represents the solution of the linear cable problem represented by a linear parabolic partial differential equation in the potential $V(x,t)$. The boundary conditions are dictated by the geometry of the neuron and the coordinatesof the locations i,j.

While solutions of this (linear) cable problem are not always easy to obtain, particular dendritic geometries yield very simple solutions. The equivalent cylinder class of dendritic geometries (Rall, 1964) is a well known example. Analytic solutions for arbitrary (branching) dendritic geometries (where each branch is described by a one-dimensional cable equation) can be generated by a graphical calculus recently developed by Butz and Cowan (1974). Their method provides the Green function $K_{ij}(\tau)$ (or its Laplace transform) for essentially all cases of interest. The membrane potential at location $x_i, V_i(t) =$ $= V(x_i,t)$ obeys then to the equation

1.1) $$V_i(t) = E_i^o + \sum_j^N K_{ij} * I_j \qquad i = 1,\ldots,M,$$

where E_i^o is the resting potential at location i, * indicates convolution and I_j is the input current at location j. V_i is linear in I_j; I_j, however, is given, for synaptic inputs, by

1.2) $$I_j = g_j(E_j - V_j),$$

where g_j is a localized (in j) conductance change and E_j is the battery associated to the corresponding ionic channel. Thus, the membrane potentials V_i are the solutions of the system of integral equations

1.3) $$V_i = E_i^o + \sum_j^N K_{ij} * [g_j(E_j - V_j)],\ i = 1,\ldots,M$$

The derivation of equation 1.3) can be generalized in the following way. The potential V(x,t) is the solution to the cable partial differential equation

1.4) $$\frac{\partial}{\partial x}\left[\frac{1}{r_a(x)}\frac{\partial V}{\partial x}\right] = c_m(x)\frac{\partial V}{\partial t} + g(x,t)V + f(x,t)\ ,$$

where $r_a(x)$ is the local axial resistance per unit length, $c_m(x)$ is the capacitance of the membrane enclosed by a unit length of cable and $[g(x,t)V + f(x,t)]$ is the non-capacitative membrane current. The solution of equation 1.4) is then given by

1.5) $$V(x,t) = \int_o^t [\int G(x,\xi;t-\tau)[g(\xi,\tau)V(\xi,\tau) + f(\xi,\tau) \ d\xi]d\tau + E^o(x)$$

where $G(x,\xi;t-\tau)$ is the Green function of

$$1.6)\qquad \frac{\partial}{\partial x}\left[\frac{1}{r_a(x)}\frac{\partial V}{\partial x}\right] = c_m \frac{\partial V}{\partial x} ,$$

with the appropriate boundary conditions. The Green function of equation 1.6) can be easily computed in several cases. Equation 1.5) reduces to equation 1.3), when the non-capacitative membrane current consists only of current through the passive membrane and through a finite number of spatially localized ionic channels, i.e. when

$$1.7)\qquad g(\xi,\tau)\, V(\xi,\tau) + f(\xi,\tau) = \frac{V(\xi,\tau)-E_o}{r_m} + \sum_{i=1}^{N} g_i(\tau)\, \delta(\xi-x_i)(V(\xi,\tau)-E_i),$$

where g_i and E_i have the same meaning as in equation 1.2), r_m is the membrane passive resistance per unit length and E_o is the resting potential.

While equation 1.3) is a vector Volterra integral equation of the second type, the more general equation 1.5) is a Fredholm integral equation (of the second kind). Properties of their solutions do not fully coincide. In the following we consider the simpler equation 1.3), which represents satisfactorily the usual physiological cases. A forthcoming paper will discuss the general problem. Let us now define the vectors $(M \geq N)$

$$1.8)\qquad \underline{V} = (V_i) \in [C(O,T)]^M , \quad \underline{g} = (g_i) \in [C(O,T)]^N ,$$

$$\underline{E}^o = (E_i^o) \in [C(O,T)]^M , \quad \underline{E} = (E_i) \in [C(O,T)]^N$$

and the matrix

$$\mathbb{K} = (K_{ij}) \in [C(O,T)]^{M,N} .$$

Furthermore we define the linear operators $\mathcal{K}$ and $\mathcal{G}$

1.9) $$\mathcal{K} : [C(0,T)]^N \to [C(0,T)]^M$$

$$[\mathcal{K} \circ \underline{h}]_i = \sum_{1}^{N}{}_j \; K_{ij} * h_j$$

$$\mathcal{G} : [C(0,T)]^M \to [C(0,T)]^M$$

$$[\mathcal{G} \circ \underline{h}]_i = (g_i \; h_i)$$

and the vector

1.10) $$\underline{F} = \mathcal{K} \circ \mathcal{G} \circ \underline{E}$$

Equation 1.3) can then be rewritten as the operator equation

1.11) $$\underline{V} = \underline{\mathcal{F}} - (\mathcal{K} \circ \mathcal{G}) \circ \underline{V} \; ,$$

where $\underline{\mathcal{F}} = \underline{E}^o + \underline{F}$.

Example 1: Lumped equivalent circuit.

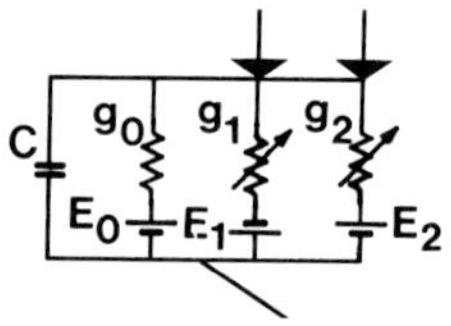

Figure 1

A small uniform patch of membrane is represented by the circuit of fig. 1. The two inputs control conductances g_1, g_2; g_o and E_o are the resting conductance and the resting potential, respectively. The circuit equation is

1.12) $$C \frac{dV}{dt} + (g_1+g_2+g_o)V = g_1 \; E_1 + g_2 \; E_2 + g_o \; E_o$$

Let us distinguish two cases.

a) The capacitance C is neglected. Then, since $K(\tau) = \frac{\delta(\tau)}{g_o}$, equation 1.3) becomes

1.13) $$V = E_o + \frac{1}{g_o}\,[g_1(E_1-V) + g_2(E_2-V)]$$

b) The capacitance C is not neglected. Equation 1.3) takes the form

1.14) $$V = E_o + \left[\frac{e^{-\frac{t}{\tau_m}}}{g_o\,\tau_m}\,u(t)\right] * \left[(g_1(E_1-V) + g_2(E_2-V))\right]$$

with $\tau_m = \frac{C}{g_o}$, since

$$K(\tau) = \frac{e^{-t/\tau_m}\,u(t)}{g_o\,\tau_m}\,,\quad u(t) = 0 \text{ for } t \leq 0,\ u(t) = 1 \text{ for } t > 0.$$

<u>Example 2:</u> One input in a dendrite.

In this case the membrane potential at a given location p is

1.15) $$V_p(t) = E_o + \int_o^t g_1(s)\,(E_1-V_p(s))\,K_{1p}(t-s)ds$$

that is, a linear Volterra integral equation of the second type. K_{1p} is the impulse response for current input at the synapse location 1 and voltage output at p.

<u>Example 3:</u> 2 inputs at locations 1,2. The potential at another location p is also desired.

Equation 1.3 takes the form

1.16) $$V_1(t) = E_o + \int_o^T g_1(s)(E_1-V_1(s))\,K_{11}(t-s)ds + \int_o^T g_2(s)(E_2-V_2(s))\,K_{12}(t-s)ds$$

$$V_2(t) = E_o + \int_o^T g_1(s)(E_1-V_1(s))\,K_{21}(t-s)ds + \int_o^T g_2(s)(E_2-V_2(s))\,K_{22}(t-s)ds$$

$$V_p(t) = E_o + \int_o^T g_1(s)(E_1-V_1(s))\; K_{p1}(t-s)ds + \int_o^T g_2(s)(E_2-V_2(s))\; K_{p2}(t-s)ds$$

2. The Neumann-Volterra Solution

In this section we proceed formally: existence, uniqueness and convergence of the solution will be discussed in the next section.

Equation 1.11) has the formal solution

2.1) $$\underline{V} = (1 + \mathcal{K} \circ \mathcal{G})^{-1}\, \underline{\mathcal{F}},$$

where $(\mathcal{K} \circ \mathcal{G})$ is a linear operator on $\underline{\mathcal{F}}$. Since $(\mathcal{K} \circ \mathcal{G})$ is causal equation 1.11) is of the Volterra type and its solution can be given in terms of the power expansion of $(1 + \mathcal{K} \circ \mathcal{G})^{-1}$, as a Neumann series

2.2) $$\underline{V} = (1 - (\mathcal{K} \circ \mathcal{G}) + (\mathcal{K} \circ \mathcal{G})^2 - \ldots)\, \underline{\mathcal{F}},$$

where the linear operator $(-\mathcal{K} \circ \mathcal{G} + (\mathcal{K} \circ \mathcal{G})^2 - \ldots)$ is the resolvent or reciprocal operator corresponding to equation 2.1). The powers of the operator $(\mathcal{K} \circ \mathcal{G})$ have kernels that are called iterated kernels, because of their structure, which is of the type, for $(\mathcal{K} \circ \mathcal{G})^p$

2.3) $$\mathbb{K}_p(s,\xi) = \int_o^T \mathbb{K}_{p-1}(s,\eta)\, \mathbb{K}_1(\eta,\xi)\, d\eta \;,$$

where $\mathbb{K}_1$ (see equation 1.3)) is the kernel of the linear operator $\mathcal{K} \cdot \mathcal{G}$. Equations as 1.3) and solutions as 2.2 are well known in the theory of linear integral equations. We are, however, interested in an explicit input- output-relation between $\underline{g}$ and $\underline{V}$. To obtain this relation we rewrite equation 2.2), as

2.4) $$\underline{V} = \underline{E}^o + (\mathcal{K} \circ \mathcal{G}) \cdot (\underline{E} - \underline{E}^o) - (\mathcal{K} \circ \mathcal{G})^2 \cdot (\underline{E} - \underline{E}^o) + \ldots \;.$$

The linear operators $(\mathcal{K} \circ \mathcal{G})^n$ are n-linear forms in $\underline{g}$. These

multilinear forms can be represented in terms of standard Volterra integrals (Palm and Poggio, 1977, Theorem 1).
Let us first observe that the (time-variant) kernels associated with the linear operators $(\mathcal{K} \circ \mathcal{G})$, $(\mathcal{K} \circ \mathcal{G})^n$, have the form

2.5) $\mathbb{K}_1(t,s) = \mathbb{K}(t-s)\ \underline{g}(s)$

$$\mathbb{K}_n(t,s) = \int_o^T \mathbb{K}_{n-1}(t-\xi)\ \mathbb{K}(\xi-s)\ \underline{g}(s)d\xi \quad .$$

Thus, the membrane potential at location i, as a functional of the inputs $\underline{g}$, can be written in terms of a multi-input polynomial series (Poggio and Reichardt, 1976; Palm and Poggio, 1977) as

2.6) $V_i(t) = h_o^i + \Sigma_j\ h_j^i * g_j + \sum_{\ell m} h_{\ell m}^i ** g_\ell\ g_m + \ldots$

where $h_{\ell m}^i ** g_\ell\ g_m = \iint_o^T h_{\ell m}^i(t-\tau_1, t-\tau_2)\ g_\ell(\tau_1)\ g_m(\tau_2)d\tau_1\ d\tau_2$

and the kernels are

2.7) $h_o^i = E_i^o$

$$h_j^i(\tau) = (E_j - E_j^o)\ K_{ij}(\tau)$$

$$h^i{}_m(\tau_1,\tau_2) = -K_{i\ell}(\tau_2-\tau_1)h_m^\ell(\tau_1) = -K_{i\ell} \ast h_m^\ell \ ,$$

where $\ast$ is the composition product which generates the "iterated kernel" of higher order. In general, the kernels have the form

2.8) $h(\tau_1, \ldots \tau_n) = -K(\tau_n - \tau_{n-1})\ h(\tau_1, \ldots, \tau_{n-1})$.

The equivalence of equation 2.6), 2.7) with equation 2.2) can be verified by substitution. Also, substitution of equations 2.6), 2.7)in equation 1.3) show that they are indeed its formal solution. The iterated structure of the kernels associated with the solution, equation 2.6), of our basic equation, equation

1.3), has various useful properties. The graphical notation of section 6 provides an easy way of writing an expansion like equation 2.6) in any specific case.

Example 1:

a) Equations 2.6) and 2.7) give

2.9) $h_o = E_o$

$$h_1 = (E_1-E_o)\,\frac{\delta(\tau)}{g_o} \qquad h_2 = (E_2-E_o)\,\frac{\delta(\tau)}{g_o}$$

$$h_{11} = -(E_1-E_o)\,\frac{\delta(\tau_1)\ \delta(\tau_2-\tau_1)}{g_o^2} \qquad h_{22} = -(E_2-E_o)\ \ldots$$

$$h_{12}+h_{21} = -(E_1+E_2-2E_o)\,\frac{\delta(t_1)\ \delta(t_2-t_1)}{g_o^2}$$

Thus

2.10) $$V(t) = E_o+(E_1-E_o)\,\frac{g_1}{g_o} +(E_2-E_o)\,\frac{g_2}{g_o} -(E_1-E_o)\left(\frac{g_1}{g_o}\right)^2 -$$

$$-(E_2-E_o)\left(\frac{g_2}{g_o}\right)^2 -(E_1+E_2-2E_o)\,\frac{g_1\ g_2}{(g_o^2)} + \ldots$$

b) One obtains

2.11) $h_o = E_o$

$$h_1(t) = (E_1-E_o)\,\frac{e^{-t/\tau_m}}{g_o\tau_m}\,u(t) \qquad h_2(t) = (E_2-E_o)\,\frac{e^{-t/\tau_m}}{g_o\tau_m}\,u(t)$$

$$h_{11}\ (t_1,t_2) = -(E_1-E_o)\,\frac{e^{-t_2/\tau_m}}{(\tau_m g_o)^2}\,u(t_2-t_1)\ u(t_1)$$

$$h_{22}(t_1,t_2) = -(E_2-E_o)\,\frac{e^{-t_2/\tau_m}}{(\tau_m g_o)^2}\,u(t_2-t_1)\ u(t_1)$$

$$h_{12}(t_1,t_2) = -(E_2-E_o)\ \frac{e^{-t_2/\tau_m}}{(\tau_m g_o)^2}\ u(t_2-t_1)\ u(t_1)$$

$$h_{21}(t_1,t_2) = -(E_1-E_o)\ \frac{e^{-t_2/\tau_m}}{(\tau_m g_o)^2}\ u(t_2-t_1)\ u(t_1)$$

...

Thus

$$2.12)\quad V(t) = E_o+h_1*g_1+h_2*g_2+K*[g_1(h_1*g_1)]+K*[g_1(h_2*g_2)]+$$
$$+K*[g_1(h_2*g_2)]+K*[g_2(h_1*g_1)]+\ldots$$

Example 2:

The kernels can be written from equation 2.7) or by substituting in equation 1.15) a Volterra series for V. One obtains

$$2.13)\quad h_o+h_1*g+h_2**gg+h_3***ggg+\ldots=E_o+K\ *g(E_1-h_o)-K*[g(h_1*g)]-$$
$$-K*[g(h_2**gg)]\ldots$$

and identification of terms of the same order gives

$$2.14)\quad h_o = E_o$$

$$h_1(t) = (E_1-E_o)K(t)$$

$$h_2(t_1,t_2) = -(E_1-E_o)\ K(t_2-t_1)\ K(t_1)$$

$$h_3(t_1,t_2,t_3) = +(E_1-E_o)\ K(t_3-t_2)\ K(t_2-t_1)\ K(t_1)$$

Thus

$$2.15)\quad V(t) = E_o+(E_1-E_o)\int_o^t K(s)\ g(t-s)ds\ -$$
$$-(E_1-E_o)\iint_o^t K(t_2-t_1)K(t_1)g(t-t_1)g(t-t_2)dt_1\ dt_2+\ldots$$

for $0 \leq t \leq T$. The graphs of section 6 provide a simple physical interpretation of such structure .

<u>Example 3:</u>

The zero order kernels are

$$2.16)\quad h_1^o = h_o^2 = h_o^3 = E_o$$

The linear kernels are

$$h_1^1(t) = (E_1-E_o)\ K_{11}(t) \qquad h_2^1(t) = (E_2-E_o)\ K_{12}(t)$$

$$h_1^2(t) = (E_1-E_o)\ K_{21}(t) \qquad h_2^2(t) = (E_2-E_o)\ K_{22}(t)$$

$$h_1^p(t) = (E_1-E_o)\ K_{p1}(t) \qquad h_2^p(t) = (E_2-E_o)\ K_{p2}(t)$$

The quadratic self-kernels are

$$h_{11}^1(t_1,t_2) = -K_{11}(t_1)\ h_1^1(t_2-t_1)$$

$$h_{22}^1 = -K_{12}(t_1)\ h_2^2(t_2-t_1)$$

...

$$h_{11}^p(t_1,t_2) = -K_{p1}(t_1)\ h_1^1(t_2-t_1) \qquad h_{22}^p = -K_{p2}(t_1)\ h_2^2(t_2-t_1)$$

The quadratic crosskernels are

$$h_{12}^1(t_1,t_2) = -K_{11}\ h_2^1(t_2-t_1)\ \ldots$$

...

$$h_{12}^p(t_1,t_2) = -K_{p1}(t_1)\ h_2^1(t_2-t_1)$$

$$h_{21}^p(t_1,t_2) = -K_{p2}\ h_1^2(t_2-t_1)$$

The membrane potential <u>V</u> is given in terms of these kernels by equation 2.6). In particular

$$2.17)\quad V_p(t) = E_o + (E_1-E_o)\ K_{p1}*g_1 + (E_2-E_o)\ K_{p2}*g_2 +$$

$$-(E_1-E_o)[(K_{p1}*K_{11})**g_1g_1] - \ldots$$

$$-(E_2-E_o)[(K_{p2}\boldsymbol{*}K_{22})**g_2g_2]$$

$$-(E_2-E_o)[(K_{p1}\boldsymbol{*}K_{12})**g_1g_2] -$$

$$-(E_1-E_o)[(K_{p2}\boldsymbol{*}K_{21})**g_2g_1] + \ldots$$

The nonlinear terms can also be written as

$$2.18)\quad -(E_1-E_o)\ K_{p1}*[g_1(K_{11}*g_1)]-(E_2-E_o)\ K_{p2}*[g_2(K_{22}*g_2)]-$$

$$-(E_2-E_o)\ K_{p1}*[g_1(K_{12}*g_2)]-(E_1-E_o)\ K_{p2}*[g_2(K_{21}*g_1)]+ \ldots$$

If location 1 coincides with location 2 and $g_1 \propto g_2$ the second order crossterms can also be written as

$$2.19)\quad -(E_2+E_1-2E_o)[(K_{p1}\boldsymbol{*}K_{11})**g_1g_2]$$

to be compared with equation 2.10). Observe again the structure of the iterated kernels. For instance

$$2.20)\quad -(E_2-E_o)[(K_{p1}\boldsymbol{*}K_{12})**g_1g_2] = -K_{p1}*[g_1(h_2^1*g_2)]$$

This term can be computed from the first order term $(h_2^1*g_2)$ simply by multiplying it with g_1 and convoluting with K_{p1}. In an analogous way all terms of order n can be obtained from terms of order n-1.

3. Convergence

In this section we consider, for simplicity of notation, the 1-input case (see example 2). Extension to the general vector case is immediate. The solution of

$$3.1)\quad V = (E^o+F) - (\mathcal{K}\circ\mathcal{G})\cdot V$$

is

$$3.2)\quad V = (1-\mathcal{K}\circ\mathcal{G} +(\mathcal{K}\circ\mathcal{G})^2-\ldots)\ (E^o+F).$$

Equation 3.2) is a Volterra series in g

$$3.3)\quad V = h_o + h_1 * g + h_2 ** gg + \ldots$$

with

3.4) $h_o = E_o$, $h_1 = (E_1-E_o)\ K(\tau)$, $h_2(t_1,t_2) =$

$= -(E_1-E_o)\ K(t_1)\ K(t_2-t_1)\ \ldots$

where K is the kernel of the linear operator $\mathcal{K}$. The convergence of series eq. 3.2) represents a well studied problem in the theory of linear integral equations. Theorems on the convergence of the Neumann series with kernels of the type $K(\tau) = \frac{L(\tau)}{\tau^\alpha}$ with $L(\tau)$ bounded in $0 \longmapsto T$ and $0 < \alpha < 1$, can be used to prove the convergence of series 3.2) (Smirnov, 1975; Petrovsky, 1971). Convergence of the Volterra series, equation 3.3), can be thus characterized through the following theorem :

Theorem

Let K be the kernel associated to the linear operator $\mathcal{K}$ of equation 3.1). Assume that

(a) K is a Volterra (causal) kernel, i.e. $K(s,\xi) = 0$ for $\xi \geq s$;

(b) K is time-invariant, i.e. $(K(s,\xi) = K(s-\xi)$;

(c) $\tau^\alpha\ K(\tau)$, $0 < \alpha < 1$ is a bounded, integrable function on $[0,T]$, $T > 0$. Then, equation 3.1) has a unique solution for all $g(t) \in C[0,T]$ and this solution, given by the Volterra series equation 3.3), is an analytic functional of g.

Proof

The n-term of equation 3.3) is a n linear form in g, $L_n g$. Its norm can be bounded by

3.5) $$||L_n\ g|| \leq ||L_n||\ \ ||g||^n .$$

Defining

3.6) $$||g|| = \sup_{0 \leq t \leq T} |g(t)| = G$$

one obtains

3.7) $$||L_n|| = \sup_{||x||=1} ||L_n\ x|| \leq \frac{N^n\ T^{n\alpha}}{n!}\ (E_1-E_o)$$

with

3.8) $$M = \Pi \sup_{0 \leq \tau \leq T} [\tau^{\alpha} K(\tau)]$$

In fact, since the kernel associated to the linear operator $\mathcal{K}$ behaves near 0 at most as $\tau^{-\alpha}$, by hypothesis, the iterated kernel associated to the linear operator $(\mathcal{K} \circ \mathcal{G})^n$ has an "iterated" kernel which behaves as $\tau^{n\alpha-1}$ near 0 (to show it one integrates equation 2.3) with $K(\tau) \approx \tau^{-\alpha}$). It follows that

3.9) $$\| L_n g \| \leq (E_1 - E_0) \frac{M^n T^{n\alpha} G^n}{n!}$$

Thus the Volterra series equation 3.4) converges absolutely and uniformly for all bounded g ($M \leq e^{MGT^{\alpha}}$) and defines an analytic functional of g (see Palm and Poggio, 1977). In the case of N inputs (see equation 2.6)), instead than one, the bound for the component V_i is

3.10) $$[\sup_{r=1,N} |E_r - E_0|] \frac{M^n T^{n\alpha} G^n}{n!} N^{n-1} \quad . \blacksquare$$

The Green function K_{ij} associated to the form of the cable problem that is used to describe passive conduction in dendritic trees, is causal, time invariant and generally continous in [0,T] with at most (when $i = j$) a pole of the type $\tau^{-1/2}$ for $\tau \to 0+$ (the Laplace transform behaves as $s^{-1/2}$). The behaviour can be seen by inspection of the general formulae of Butz and Cowan and depends on the existence of a well behaved Laplace transform of the solution of the cable problem. Thus, the theorem implies that the membrane potential that results from arbitrary, bounded conductance changes $\underline{g} \in C[0,T]$ in a passive, cable-like dendritic tree is an analytic functional of g.

Examples

Examples 1b, 2, 3 satisfy the conditions ot the theorem. The Volterra series (2.12), (2.15) and (2.17) represent uniquely the solution and converge uniformly for all bounded inputs g in [O,T]. The Green function associated with example 1a does not satisfy the conditions of the theorem; the solution, equation 2.10), does not, in fact, converge everywhere but only for $\left|\frac{g_1+g_2}{g_0}\right| < 1$. The solution, which is still a Volterra-like series where the kernels are distributions (Palm and Poggio, 1977), can be continued analytically. The reason for the anomaly of example 1a is due to the degenerate nature of the associated linear problem (the parabolic equation is reduced to an algebraic equation).

4. Underlying Hypotheses and Possible Applications

The physiological hypotheses underlying our approach are the usual ones (Jack, Noble and Tsien, 1975; Rall, 1964), namely:

(a) The unit of membrane area is represented as a resistance in parallel with a capacitance. The resistance and the capacitance are voltage independent (passive). The associated batteries do not change during membrane conductance changes.

(b) To obtain equation 1.3) each synaptic conductance change is assumed to be localized in a negligible area of the membrane and does not change the passive properties of the membrane. This is not required in the derivation of the more general equation 1.5).

Other hypotheses as equipotentiality of the nerve membrane, homogeneity, isotropy etc. or even stronger ones as the equivalent cylinder hypothesis of Rall simplify the linear problem of obtaining the Green function of equation 1.3), but do not affect the nonlinear approach. Two remarks are important:

1) We have considered here the transduction conductance changes (at the level of the postsynaptic membrane) $\rightarrow$ voltage (at some location in the cell). A complete description of synaptic transduction requires, however, also consideration of the transformation of the presynaptic potential into postsynaptic conductance change. A forthcoming paper will deal with the full problem.

2) The general nonlinear problem (equations 1.1) and 1.2)) is reduced in a <u>standard</u> way to the linear problem of finding the Green function of the cable equation. This problem, although quite difficult in some cases, is much studied and a "catalog" of solutions already exists (Rinzel and Rall, 1974; Jack, Noble and Tsien, 1975; and the electric cable literature). In addition, the graphical calculus developed by Butz and Cowan provides analytic solutions for the transfer function of the cable equation at any point on the dendritic tree of neurons with arbitrary branching geometries. Solutions for simpler geometries may also be satisfactory in many problems. Approximations of the Green function (for instance for large or small t) may often yield useful, simple expressions. Estimates of the Green function can also be provided by numerical methods (see Rall, 1964 for his compartmental analysis).

There are several applications of the methods outlined in this paper. Let us mention some of the most interesting ones:

a) It becomes possible to deal with the case of arbitrarily located conductance inputs in arbitrary neuron geometries. For instance, various properties of the nonlinear interaction between two synaptic inputs, studied numerically by Rall and others, can be recovered from an examination of the first terms of equation 2.17). The 2 quadratic terms in g_1 and g_2 provide nonlinear summation reducing the effectiveness of the inputs; they depend in a characteristic way from the associated ionic battery and from both the local cable properties and the location of the input with respect to the location p for the membrane potential V_p. In particular it is clear (see especially equation 2.17)) that changes in g_1 of duration much smaller than the "duration" of K_{11} will give a negligible nonlinear contribution (since the product $g_1(K_{11}*g_1)$ will be different from zero only when g_1 is different from zero, whereas the linear term will be different from zero for the whole "duration" of K_{11}). The crossterms give the nonlinear interaction between the two conductance changes: timing and duration depend on the geometry and on membrane properties. If the two conductance changes occur at the same site ($1 \equiv 2$), simultaneous conductance changes of sufficient duration provide the maximum nonlinear interaction. For non adjacent locations the maximum nonlinearity occurs when the sum of the overlap between g_2 and $(K_{21}*g_1)$ and the overlap between g_1 and $(K_{12}*g_2)$ is maximum. Especially interesting is the case in which g_1 is an excitatory channel $(E_1 >> E_o)$ and g_2 is inhibitory with a reversal potential near the resting membrane potential $(E_2 \sim E_o)$. In this case the terms containing $(E_2 - E_o)$ in equation 2.16) and 2.17) can be neglected and the first term in which g_2 appears is the crossterm $-(E_1 - E_o)\ K_{p2}*[g_2(K_{21}*g_1)]$. Several general properties of post-synaptic inhibition, as the dependence of its effects on its dendritic or somatic location, can be directly derived from this term (compare Jack, Noble, and Tsien, 1975, p. 197). It is worth mentioning that the discussion of specific cases are greatly facilitated by the use of the graphs introduced in the next sections.

b) The methods developed in this paper seem to provide useful tools to deal with diades and reciprocal synapses or more complex synaptic arrangements, where the elementary computing components may be as small as dendritic membrane patches. Here a critical property of a Volterra type representation like equation 2.6) is that it allows to put systems or components together in standard ways, exactly as the impulse response or transfer function approach in the case of linear systems (see Poggio and Torre, 1977).

c) The Volterra representation of the membrane potential offers a rather transparent link with the information processing properties of the transformation implemented by a particular synaptic geometry. We will mention further this point in section 6.

5. Fourier Transforms and Symmetry Properties of the Kernels

a) It is usually easier to solve the linear cable equation in terms of its transfer function $\tilde{K}(\omega)$, that is the Fourier or Laplace transform of the Green function $K(\tau)$. It is therefore, interesting to consider the Fourier transform of the solution equation 2.6), that is (for the one input case)

$$5.1)\qquad \tilde{V}(\omega) = H_o\ \delta(\omega) + H(\omega)\ G(\omega)$$

$$+\ H_2(\omega_1,\omega-\omega_1)\ G(\omega_1)\ G(\omega-\omega_1)d\omega_1 + \ldots,$$

where the iterated kernels are given by

$$5.2)\qquad H_o = E_o$$

$$H_1(\omega) = (E_1-E_o)\ \tilde{K}(\omega)$$

$$H_2(\omega_{11},\omega_2) = -\tilde{K}(\omega_2)\ H_1(\omega_1+\omega_2)$$

$$H_3(\omega_1,\omega_2,\omega_3) = -K(\omega_3)\ H_2(\omega_3+\omega_2,\omega_1)$$

The structure of the kernels is simple also in the frequency domain. Each kernel can be obtained from the lower order one and the linear transfer function $\tilde{K}$. The specific structure of the kernels implies that $\tilde{V}(o) = \overline{V(t)}$ consists of the sum of terms always containing the term $\tilde{K}(o)$ as a multiplicative factor.

b) Kernels as the ones given by equation 2.7) are not symmetric. The selfkernels can be of course symmetrized as usual (Bedrosian and Rice, 1971; Halme, 1971). The crosskernels involving two or more inputs can be decomposed into the invariant representation of a permutation group. For instance, in the two input case (equation 2.16)) one can decompose the cross-kernel sum $(h^p_{21} + h^p_{12})$ into a symmetric and an antisymmetric component; specific functional properties may be associated to the two components (Geiger and Poggio, 1975).

6. Graphs and Interactions

There is a superficial analogy between the series (2.2) and the S matrix expansion of quantum field theory: both are Neumann series, solutions of linear integral equations; the linear kernel is in one case the Green function $K(\tau)$, in the other case the interaction Hamiltonian; in both cases the iterated kernels of second and higher order are constructed in a similar way from the linear kernel. This analogy suggests that a graphic notation somewhat similar to Feynmann diagrams may be used to "write" symbolically expansions like, for instance, equation 2.16). Consider, for simplicity, the expansion of V_p equation 2.16) and 2.17): the linear term is given by a linear operator which "propagates" the input g_1 in 1 to p. The second order term in g_1 consists of the linear propagation of input 1 to 1, multiplied by g_1 and then propagated to p. Similarly the mixed term in g_2, g_1 is given by the propagator from 1 to 2, multiplied by g_2, propagated then from 2 to p.

A possible graphical notation is based on the following two elementary graphs

1)  $\equiv K_{ji} *$, i.e. at j one finds $(K_{ji} * g_i)$ if g_i is the input in i.

2)  ≡ multiplication between input at m and output of "propagator"  ; the effect is at location m.

Three additional conventions must be respected:

3) the order n of the term is the number of the symbols of type 2 plus one. The sign of the term is $(-1)^{n-1}$.

4) The whole term corresponding to a graph must be multiplied by the factor $(E_i - E_o)$, where i is the starting location of the first propagator of type 1.

5) The zero-order term E_p^o does not have a corresponding graph.

With these rules equation 2.16) can be read from the following graphic notation:

6.1) $V_p = E_p^o +$ 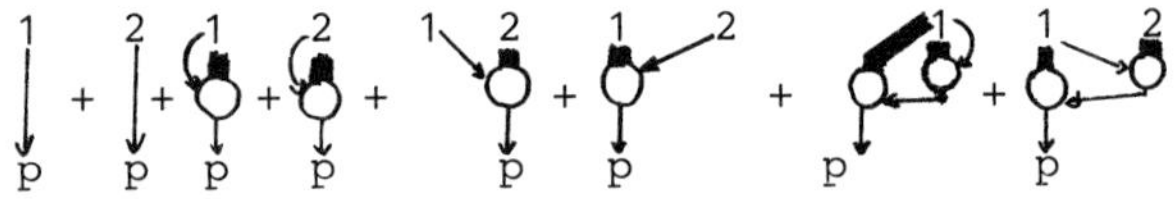 $+ \dots$

The graphical notation used here is not the same notation introduced (Poggio and Reichardt, 1976), for a general Volterra series (although it is consistent with it) since the kernels have here a specific structure. Alternatively one can also use the notation 1⟶②⟶①⟶p to denote, for instance, the last term of equation 6.1) and similarly with other terms.

In equation 6.1) the basic linear graph is ↓ , the basic second order graph is of the type  , the third order graphs are built from the second order ones substituting the first with a second order graph, and similarly for higher order interactions. It is also easy to read equation 2.16) from the graphic notation and the rules 1-5. For instance, the graph

leads immediately to $-K_{32}*[(K_{21}*g_1)g_2](E_1-E_0)$, where one starts to write (E_1-E_0), then $K_{21}*g_1$ and so on, following the structure of the graph. This graphic notation may facilitate the interpretation of the various terms in an expansion like equation 2.6) for a specific problem. In addition, it offers also a standard way of sorting out the important terms for a particular problem. Thus, given a neuronal geometry with some synaptic inputs, it is easy to write down the relevant graphs (see the example of fig. 2). If the various K_{ij} required are available (for this the corresponding linear problems must be solved with the given geometry) at least approximatively, together with the ionic batteries, it may become relatively easy to interpret the various graphs and the interplay of the various conductance inputs.

Interestingly, the graphs have a strong intuitive meaning and they could have obtained almost without mathematics, simply from the physics of the problem. Consider, for instance, the effect at location p of a conductance change in location 1: ΔV_p will depend on the propagation to p of the current at 1. The current itself is given by g_1 time the local driving potential which is E_1-E_0 corrected by the effect of the current on E_0 and so on. In some sense the graphs make explicit the series of physical processes which determine ΔV_p. Since the propaga-

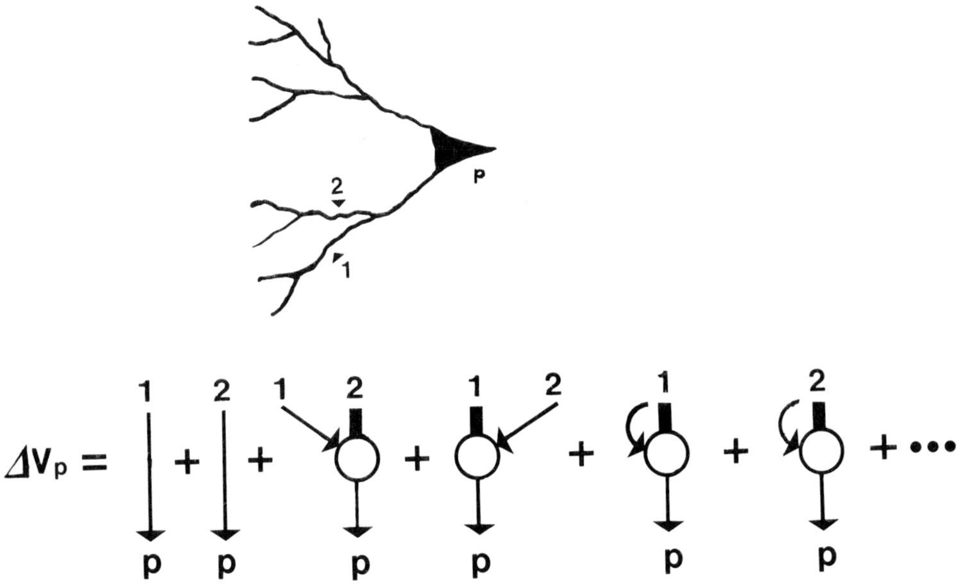

Figure 2

tion of the effects is not instantaneous the various graphs (or terms of the expansion) actually represent successive, "distinct" physical processes.

Estimates of errors arising by the truncation of the series are also easy (see section 3). The convergence of the series can be improved by the so called Euler transformation, especially devised for alternating series as equation 2.17) or 6.1). Let us consider, for simplicity, the one dimensional equation 2.15). Euler transformation gives the series

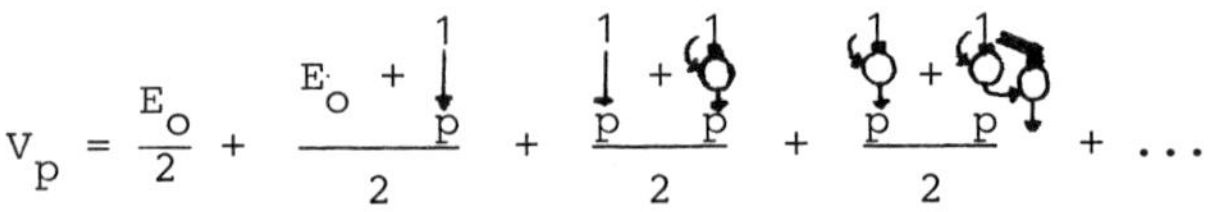

This implies that if this series is truncated at the n-th order, the n-th order term of series 2.15 contributes here with a factor 1/2, whereas the sum of the previous terms coincides.

One more remark is worthwhile. From the system identification point of view, knowledge of the Volterra kernels up to the second order allows, in principle, the complete diagnosis of such "synaptic" systems (equation 1.3)) with N inputs. All other terms of the series are then uniquely determined, i.e. two synaptic systems are identical if and only if their Volterra kernels up to the second order coincide (identification up to the quadratic kernels can be carried out, in principle, with delta pulses in the conductance inputs). Another line of reasoning also seems to support the conjecture that the "qualitative" properties of such synaptic systems are contained in the linear and quadratic terms of the Volterra expansion. Consider a 2-input system: since the Volterra representation of ΔV_p converges absolutely and uniformly and is an alternating series, the linear and the quadratic approximations represent upper and lower bounds of the solution $\Delta V_p(t)$; the quadratic approximation is, moreover, the first one that always contains both input signals g_1 and g_2, independently from the values of E_1, E_2 (see the example equation 7.3)).

7. Information Processing, Elementary Computations and Synaptic Morphologies

It has been proposed (Geiger and Poggio, 1975) that a Volterra-like representation of a smooth functional or system (Palm and Poggio, 1977) can provide an insight into its information processing properties. Specific computational properties can be, in fact, associated with Volterra terms of a given order and type. A Volterra representation can be used to classify in a canonical way the information processing of "simple" systems or algorithms (Poggio and Reichardt, 1976; Marr and Poggio, 1976). For instance, it has been shown that computation of movement requires at least a second-order interaction between pairs of inputs and this minimal interaction is actually implemented in the visual system of the fly. Relative movement detection requires interactions of at least fourth order between four input regions. As a consequence, the Volterra

form of our solution equation 2.6) is an interesting fact. Three points are important. Firstly, there becomes available, as we mentioned earlier, a rather transparent link between a given specific "local circuit", which can be described by equation 1.3), and its information processing properties. The second point rests on the demonstration that the membrane potential at an arbitrary location is an analytic functional of bounded conductance inputs (theorem 1). This implies that the Volterra representation equation 2.6) is here very natural and in some sense intrinsic to the nature of the transduction. In addition, if local circuits, as represented by equation 2.6), are the elementary computing elements of nervous systems, the use of the Volterra formalism to classify and describe simple information processing by nervous systems will receive an interesting support. The third point consists of the new possibility of asking directly for the possible neural implementation of a given Volterra interaction. For instance, since movement detection is known to depend on a second-order Volterra interaction between two inputs, one can ask which local circuits may implement it. The answer we proposed (Torre and Poggio, 1978) is that a diadic synapse performs the required multiplication-like interaction.

In the following we want to outline in a very simple way a few examples of how elementary computations can be connected with specific local circuits, i.e. synaptic morphologies.

A synapse of the diade type can be modelled through the circuits shown in fig. 1, which is the electrical equivalent of a patch of dendritic membrane receiving two distinct inputs, through two distinct, closely adjacent synapses. The circuits equation is equation 1.12) or more generally equation 1.16). The solution, equation 2.17), reveals, for $E_1 \gg E_o$ and $E_2 \sim E_o$ that changes in the somatic potential V_p depend on g_1 and g_2 as

7.1) $\Delta V_p = (E_1 - E_o)\{K_{p1}*g_1 - K_{p1}*[(K_{11}*g_1)g_1] - K_{p2}*[(K_{21}*g_1)g_2]\}$

and, if $g_2 \gg g_1$,

7.2) $\Delta V_p \simeq (E_1 - E_o)\ K_{p1} * g_1 - (E_1 - E_o)\ K_{p2} * [(K_{21}\ g_1) g_2]$

which can be directly read from

7.3) $\Delta V_p =$ 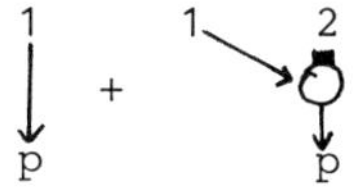 ,

since the other graphs (up to the second order) are negligible here. If one considers the approximation equation 1.7), equation 7.2) simplifies to

7.4) $$\Delta V_p = (E_1 - E_o)\ \frac{g_1}{g_o} - (E_1 - E_o)\ \frac{g_1\ g_2}{g_o^2}$$

which is discussed elsewhere (Torre and Poggio, 1978). Fig. 3a shows a diadic synapse (the equivalent circuit is given in fig. 1) which, with the appropriate parameter values, implements the interaction equation 7.4). Thus, the interaction is roughly of the type $y = x_1 - \alpha\ x_1 \cdot x_2$; fig. 3b suggests how, in principle,

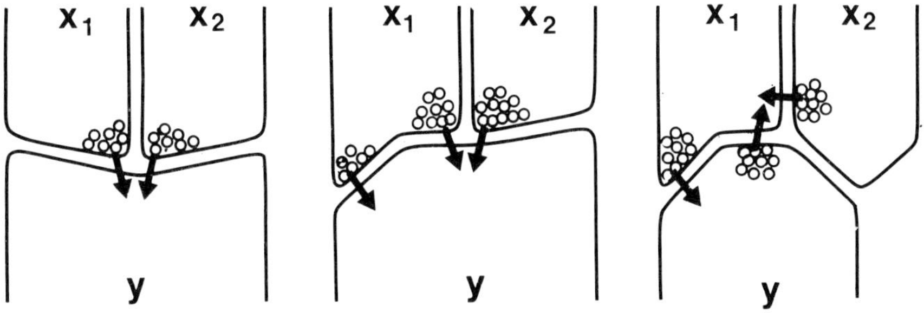

Figure 3a,b,c

one could obtain a "pure" multiplication. The added synapse is inhibitory and if its distance from the diade is in the order of the length constant of the membrane, its contribution adds algebraically, eventually cancelling the linear term of equation 7.2) (fig. 3b). The "diade" provides also a rather natural way of dividing 2 signals, through the reciprocal arrangement of fig. 3c. Assuming again that the diade implements in fig. 3c the operation $y - \alpha y x_2$ one obtains, formally, if

$y = \beta x_1$

and

$x_1 = \tilde{x}_1 + y - \alpha y \tilde{x}_2$,

where $\tilde{x}_1$ and $\tilde{x}_2$ are external input, the relation

$$y = \beta \frac{\tilde{x}_1}{1-\beta+\tilde{x}_2} .$$

If $\beta \simeq 1$ the result is

$$y = \frac{1}{\alpha} \frac{\tilde{x}_1}{\tilde{x}_2} .$$

Clearly fig. 3b and 3c (and similar ones) are oversimplified schemes. But they illustrate how local circuits, whose components are highly localized regions of the dendritic membrane, may implement elementary, specific interactions and computations through special junctional structures. In depth mathematical analysis of local circuits of this type can be carried out with the tools discussed in the first part of this paper (Schmidt el al., 1976).

Conclusions

We believe that the approach outlined in this paper may provide a theoretical background for analyzing the so called "local circuits" and for establishing a link with the information processing performed by a nervous system. The approach is clearly at the "hardware" level (compare Marr and Poggio, 1976) since it deals with basic components like synapses and dendrites. It provides, however, through the Volterra representation, an helpful, but of course not univocal, link with the higher level of understanding a nervous system, at which one considers how a computation is carried out and tries to classify and describe the algorithms involved.

The theory outlined in this paper is still at an embryonic stage and has to be extended in several directions. We feel, however, that, together with our earlier analysis of the neural encoding process (Poggio and Torre, 1977), it may represent a starting point for an understanding of neural circuits and their role in elementary processing of information.

Acknowledgments

We are grateful to A. Borsellino, J.D. Cowan, W. Reichardt, C. Wehrhahn and especially G. Palm for several useful discussions. We thank L. Heimburger who draw the figures and I. Geiss who cared about the English manuscript.

References

Bedrosian, E., Rice, S.O.: Proc. IEEE 59, 1688 (1971).
Blomfield, S.: Brain Res. 69, 115-124 (1974).
Butz, B.G., Cowan, J.: Biophys. J. 14, 661-689 (1974).
Geiger, G., Poggio, T.: Biol. Cyb. 17, 1-16 (1975).
Halme, A., Orava, J., Blomberg, H.: Int. J. Systems 2, 25 (1971).
Kuffler, S.W., Nicholls, J.G.: "From neuron to brain" Sinauer Ass. (1977).
Jack, J.J.B., Noble, D., Tsien, R.W.: Electric Current Flow in Excitable Cells. Clarendon Press, Oxford (1975).
Marr, D., Poggio, T.: M.I.T. Memo 357 or Neurosciences Research Program, Bull., eds. Dowling, Held, Pöppel (1976).
Palm, G., Poggio, T.: SIAM J. on Appl. Math. 33, 2 (1977).
Petrovsky: "Lectures on the theory of integral equation" MIR Moscow (1971).
Poggio, T., Reichardt, W.: Quart. Rev. Biophys. 9, 377-438 (1976).
Poggio, T., Torre, V.: Biol. Cyb. 27, 113-124 (1977).
Rall, W.: In: Neural Theory and Modeling. R.F. Reiss, ed., Stanford University Press (1964).
Rinzel, J., Rall, W.: Biophys. J. 14, 759-790 (1974).
Schmitt, F.O., Parvati Dev, Smith, B.H.: Science 193, 114-120 (1976).
Smirnov, V.: Cours de Mathèmatiques Superieures. MIR Moscow (1975).
Torre, V., Poggio, T.: Preprint (1977).
Whittaker, E.T., Watson, G.N.: "Modern Analysis". Cambridge University Press (1952).

FIGURE-GROUND DISCRIMINATION BY THE VISUAL SYSTEM OF THE FLY

Werner Reichardt
Max-Planck-Institut für biologische Kybernetik Tübingen

Introduction

A small black object embedded in a black-white contrasted random-dot texture, is easily confused by a human observer with the ground. However, even small relative motions between object and ground allow an easy detection. A similar observation is made when the object is replaced by a figure consisting of the same random-dot texture as the ground. As long as the figure is not moved against the ground, figure-ground discrimination is not possible as the boundary of the figure disappears in the texture of the ground. However, when the figure is moved relative to the ground, the boundary of the figure can be easily seen. In some sense relative motion between figure and ground leads to an independent or context free perception of a figure. The observations are of course different when the texture of the figure differs from the texture of the ground. Under these conditions one can usually see the figure's boundary and its content even if it is not moved relative to the ground. However, the impression one receives from such a figure seems different whenever it is moved, an observation which is again in accordance with the hypothesis that relative motion between figure and ground leads to an independent or context (ground) free perception of the figure.

In recent years we have analyzed the visual orientation behaviour in the fly. Flight behaviours of houseflies (*Musca domestica*) demonstrate elaborate visual control systems. Flies perceive motion relative to the environment and thereby stabilize their flight course; they locate and fly towards objects; they track moving targets and chase other flies; they discriminate or prefer specific visual patterns; they even discriminate a texture into object or into figure and ground. (See the review articles by Reichardt and Poggio, 1976 and by Poggio and Reichardt, 1976.)

In this paper I will especially deal with the computations underlying object-ground and figure-ground discrimination through relative movement carried out by the visual system of the fly.

Experimental Procedures

In free flight, an insect possesses six dynamic degrees of freedom (neglecting head movements): three of translatory and three of rotatory motion. The investigations undertaken so far have each been confined to either one degree of rotation (the rotatory motion of a horizontally flying fly around its vertical axis; Reichardt, 1973) or one degree of translation (the vertical motion of a horizontally flying fly; Wehrhahn, 1974; Wehrhahn and Reichardt, 1975). Measurements were carried out by means of highly sensitive, fast mechanoelectric servo-transducers, which fix the fly in space and either sense the flight torque, or the lift force generated by the wings of the test fly.

When a contrasted optical environment is moved or flickered in front of a fixed flying test fly and the transducer signal recorded, one is operating under "open-loop" conditions. When, however, the transducer signal is used to control the position and speed of the environment through a simulation of the flight dynamics, the conditions are "closed-loop". Contrasted environments were either provided by patterns mounted on cylinders or by electronically generated patterns displayed by oscillographic monitors.

The experiments were carried out with female *Musca domestica*, head fixed to the thorax.

Open-loop and Closed-loop Behaviour

A phenomenological theory (Poggio and Reichardt, 1973a),

which takes into account the dynamics of flight, links (open-loop) information processing in the visual system of the fly with the natural orientation behaviour. It has been shown that (closed-loop) orientation behaviour of the fly can be quantitatively predicted from knowledge of the open-loop response. This means that one can consider the computations performed on the visual input by the nervous system of the fly as independent from the motor loop being "open" or "closed". In addition, it suggests that an analysis of the computations performed under open-loop conditions, as it is very often the case, is in fact completely sufficient for an understanding of the behaviour.

In this connection it should be mentioned that the "Reafferenz-Prinzip" (von Holst and Mittelstaedt, 1950) seems not to be necessarily required at the level of the behaviour studied so far in flies. We have not observed any example of closed-loop orientation behaviour, which can not be accounted for in terms of open-loop responses.

Movement and Position Computations

Flies can compute relative motions and positions of objects in the optical environment. In order to make the basic logics of these computations understandable, I will discuss two open-loop experimental pradigms: Let us assume that a one dimensional periodic grating (sinusoidal contrast) is moved horizontally at constant speed in one and then in the opposite direction, in front of two adjacent photoreceptor. This experiment has been performed by Kirschfeld (1972) through optical stimulation of specific pairs of photoreceptors in one ommatidium of the fly's compound eyes. The light stimuli which represent the inputs to the two receptors are sinusoidally modulated with the same frequency, the same amplitude and a phase shift whose sign reflects the direction of movement. The experimental result shows that the fly responds with a strong direction-sensitive average optomotor response, which changes sign if the direction of motion is inverted. It is obvious that a necessary condition for the evaluation of directed movement is a system with at least two inputs which performs a computation that cannot be carried out by means of only linear components. The underlying overall interaction between the two input signals must be nonlinear, since the average output of a linear system depends on the mean values of the inputs and not on their phase relationship. A detailed experimental and theoretical analysis (see footnote) has lead to the conclusion that the interaction is a second-order nonlinearity. The following integral representation describes the input-output relation of this nonlinear interaction with x_ν and x_μ being the input functions to the receptors ν and μ

$$(1) \qquad y_{\nu\mu} = \iint_{-\infty}^{+\infty} h_{\nu\mu}(\tau_1,\tau_2)\, x_\nu(t-\tau_1)\, x_\mu(t-\tau_2) d\tau_1\, d\tau_2$$

where $h_{\nu\mu}$ is a kernel with the antisymmetry property $h_{\nu\mu}(\tau_1,\tau_2) = -h_{\nu\mu}(\tau_2,\tau_1)$ and $y_{\nu\mu}$ the associated behavioural (torque) response component of the fly. The integral representation describes the input-output relation of an antisymmetrized elementary movement direction sensitive evaluation system in the fly. A graphical representation of this interaction is shown in Fig. 1a. Experimental evidence suggests that the direction sensitive torque response does not depend upon the location of the two interacting receptors in the compound eye.

In the second experimental paradigm a narrow stripe is flickered in front of one eye of the fly at a given angular position ψ_o. Again, the light stimuli onto the receptors near

---

Hassenstein, 1951; Hassenstein and Reichardt, 1956; Reichardt, 1957; Hassenstein, 1958, Hassenstein, 1959; Reichardt and Varjú, 1959; Varjú, 1959; Reichardt, 1961; Fermi and Reichardt, 1963; Götz, 1964; McCann and MacGinitie, 1965; Thorson, 1966a and 1966b; Hengstenberg and Götz, 1967; Varjú and Reichardt, 1967; Reichardt, 1969; Reichardt, 1970; Götz, 1972; Eckert, 1973; Marmarelis and McCann, 1973; Poggio and Reichardt, 1973a and 1973b; Buchner, 1974; Pick, 1974a, 1974b and 1976; Poggio 1974; Geiger and Poggio, 1975; Götz, 1975, Poggio, 1975; Buchner, 1976; Pick and Buchner, 1978.

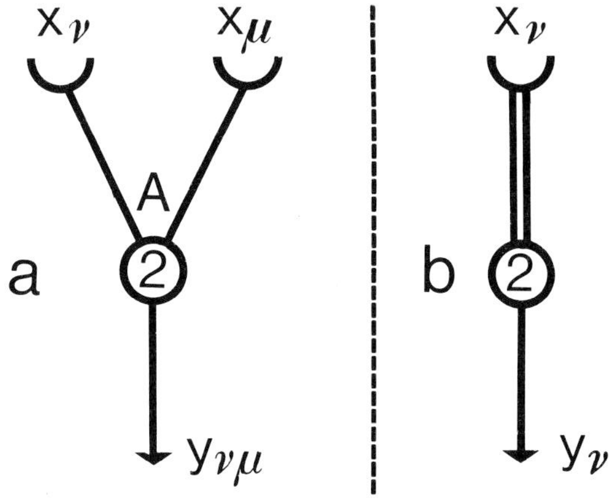

Fig. 1: Graphical representations of the interactions described by equations (1) and (2).
a) Antisymmetric (A), second-order crossinteraction of the input functions $x_\nu(t)$ and $x_\mu(t)$. Due to the antisymmetry the response, $y_{\nu\mu}$ changes sign when the input functions are exchanged (reversal of the direction of object motion).
b) Second-order self-interaction of the input function x_ν. The interaction is symmetric.

that position are sinusoidally modulated in time, with an identical frequency and phase. A similar stimulus is generated by a stripe oscillating around that position with a small amplitude. Both types of stimulations elicit an average tendency of the fly to turn towards the stimulus. The average response is direction insensitive, by definition. Again, no linear system can perform this computation. The reason is that in the time average of the response no significant reaction is found for a stabilized retinal image. Receptor input modulation is necessary to elicit a direction-insensitive response. Hence, the operation on the inputs cannot be linear, since the average output of a linear system is independent of the input modulation. Various experiments and their analysis (Reichardt, 1973; Poggio and Reichardt, 1973a; Poggio and Reichardt, 1973b; Pick, 1974a;

Geiger and Poggio, 1975) suggest that position information of rather small objects is computed by second-order self-interaction of optical information received from single receptor inputs. The input- output-relation between one receptor x_ν and the measured torque response component is approximated by the integral representation.

$$(2) \qquad y_\nu = \iint_{-\infty}^{+\infty} g_\nu(\tau_1,\tau_2)\ x_\nu(t-\tau_1)\ x_\nu(t-\tau_2)d\tau_1\ d\tau_2$$

whose graphical representation is given in Fig. 1b.

From these considerations it follows that spatially distributed nonlinear functional interactions between channels following the receptor outputs are essential for the computa-

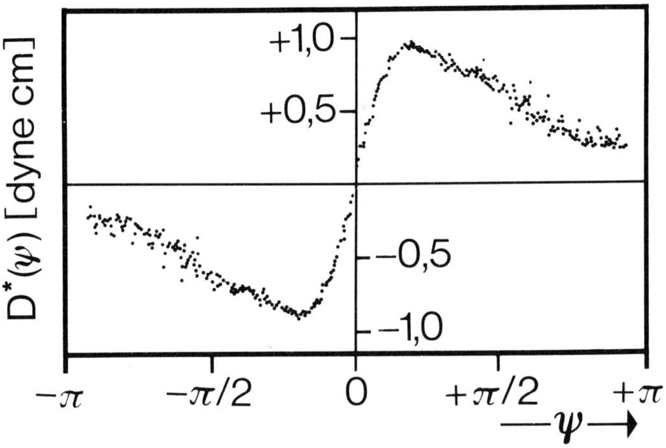

Fig. 2: The (direction-insensitive) torque response component D^* induced by a vertical black stripe (5° wide). The function D^* represents the mean attraction of the fly towards the stripe, for each ψ. To avoid a stabilized retinal image, to which the fly would not response, the stripe was randomly oscillated by the fly around each ψ position with small amplitude (partial closed-loop conditions, see Reichardt, 1973). The record is an average taken from measurements with 111 test flies. Redrawn from Reichardt, 1973.

tions of movement and position information in the fly's visual system. The computations of movement do not depend upon the location on the eye whereas the computations of position depend on the location. Fig. 2 shows how the average responses to the position computations depend on the angle ψ which designates the angular position of the image of an object (stripe) on the two compound eyes. $\psi = 0$ refers to the vertically oriented symmetry line between the two compound eyes.

Lateral, Nonlinear, Inhibitory Interactions

A first hint that inhibitory lateral interactions affect the position computations came from experiments with two objects (vertical stripes); Reichardt, 1973; Reichardt and Poggio, 1975. A pattern composed of two vertically oriented stripes leads to an attraction of the fly which turned out to be less than the sum of the attractions to each of the two individual stripes.

In a similar way, the attractiveness of a flickered vertical stripe (see Fig. 3, filled dots) does not increase proportionally to its width but even decreases when its lateral dimensions exceed about 10°; that is five columns of adjacent ommatidia in the fly's compound eye. Thus, the emerging organization seems to include self-interactions and surrounding lateral interactions, on an angular range larger than $\sim 10^{\circ}$ but certainly less than $\sim 80^{\circ}$. Fig. 3 (square dots) also shows a comparison between the time averaged response of the fly to a stimulus consisting of a black stripe oscillated sinusoidally with different amplitudes A and a flicker signal of the same frequency. In the first case (oscillating stripe) mainly the self-interactions are stimulated. Consequently the reaction measured builds up roughly linearly with the oscillation amplitude A whereas the response to a flickering bar decreases beyond some width presumably because the lateral interactions come into play (see also Pick, 1976).

An important question for an understanding of the position computations concerns the order of the lateral interactions and, possibly, their spatial organization. A first answer to this

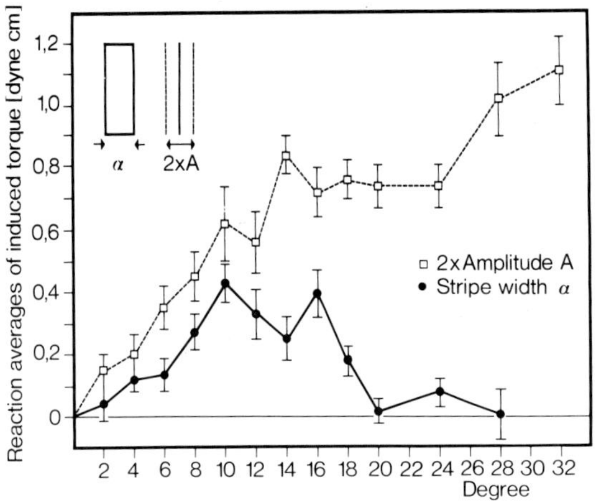

Fig. 3: Average, open-loop responses from test flies to a stationary, flickered vertical stripe as a function of its width d(●) and to a black stripe oscillated sinusoidally with an amplitude A(2,5 Hz frequency) (□). The data refer to a mean object position $\psi_o = \pm 30^o$.

The average brightness of the flickered stripe amounted to 468 cd/m^2, the brightness of the background to 64 cd/m^2. Modulation degree was m = 88%. The stripe was oscillated with a background of 700 cd/m^2. Each point in both experiments is the mean of 30 measurements with 15 individual flies. Each measurement lasted 2 min. The vertical bars denote standard errors of the mean. A similar experiment has been carried out by Pick, 1976. The illumination conditions in his experiment were different.

problem was again provided by two-input experiments (Pick, 1974a). Two adjacent vertical dark stripes were sinusoidally flickered on the fly's right eye, with various phase shifts. The results (Fig. 4) require interactions of degree 4 or higher. The geometry and the number of interacting photoreceptors were as yet not clear. At least two receptors, and perhaps more, are required. In Fig. 5 two of the most simple alternative, given by graphical representations of interactions, are shown. The simple organization shown in Fig. 5a envisages excitatory second-order

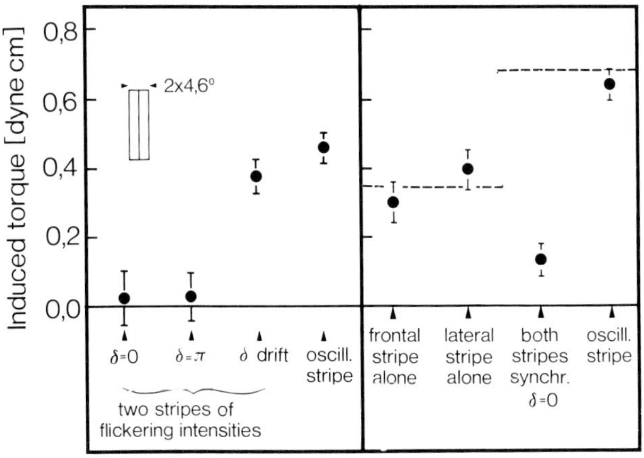

Fig. 4: Average torque response to different stimuli of sinusoidally flickering and oscillating stripes. Mean angular position of the stimulus is $\psi_o = \pm 20^o$. The data are from six (left) and eight (right) female flies, respectively. The flicker and oscillation frequency amounted to 3 Hz, the amplitude of the oscillation to $A = 4.6^o$. No attraction is observed towards the two stripes flickering simultaneously ($\delta = \pi$). However, flicker of the two stripes with two slightly different frequencies (3 Hz and 3.25 Hz) elicit a mean attraction which is in the same order of the response elicited by the oscillating stripe. The attractiveness of an oscillating stripe is almost the sum of the attractivenesses of the single flickering stripes (right figure). Redrawn from Pick (1974b).

self-interactions and lateral inhibitory fourth-order interactions with various spacings. The number of receptors per elementary interaction configuration is two. Another possibility is given in Fig. 5b where the interactive structure consists of four receptor inputs.

Object-Ground and Figure-Ground Discrimination

Under experimental and natural (closed-loop) conditions, flies can fixate and track small objects in front of a random texture, if the object moves relative to the ground (Virsik

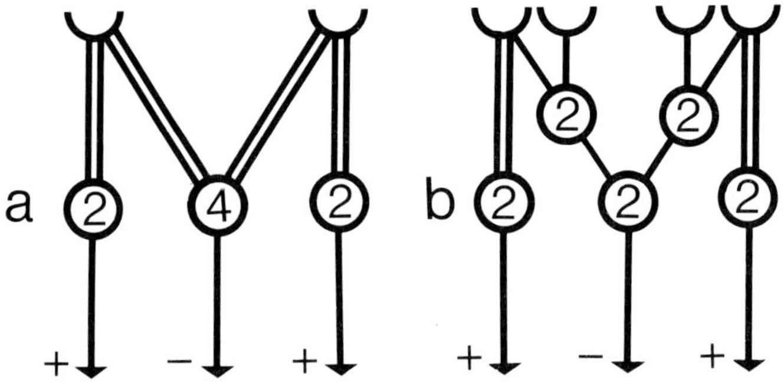

Fig. 5: Two interactive organizations which could underlie the object-ground and the figure-ground discrimination effect in the fly. The interaction schemes contain second-order self-interactions and fourth-order cross-interactions with two (a) and four (b) receptor inputs. The numbers in the circles designate the order of the interaction.

and Reichardt, 1976). Open-loop experiments also demonstrate the same effect.

It is quite clear that computations which do not include lateral interactions between different receptor inputs, cannot account for this effect. However, it is possible to formally prove, that in principle the simplest interaction graphs capable of this computation have order four and two receptor inputs.

To find out whether in flies this computation can be actually accounted for by the "minimal" fourth-order graph of Fig. 5a a series of open-loop experiments have been performed (Heimburger, Poggio and Reichardt, 1976; Poggio and Reichardt, 1976). The inset of Fig. 6 shows the basic experimental design. A black vertically oriented object (stripe) is sinusoidally oscillated around a fixed position in front of one eye. A random-dot ground can be also oscillated with preset amplitude A, fre-

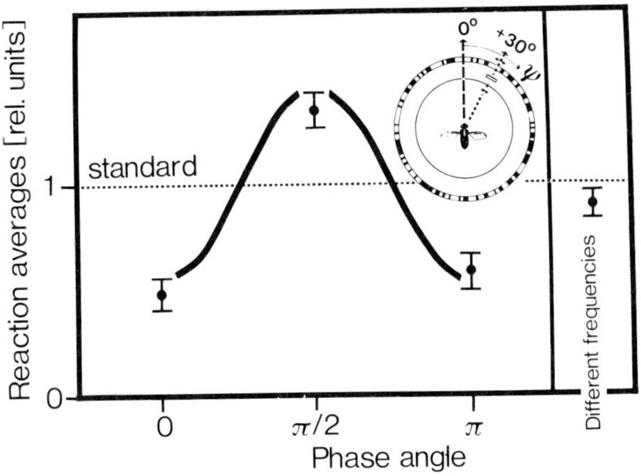

Fig. 6: Average torque response of ten flies to sinusoidally oscillating object and ground patterns, under open-loop conditions. The object consists of a black, vertical stripe, 3° wide, positioned in the lower part of the panorama oscillated around the mean positions $\psi_o = \pm 30^{\circ}$.

The ground pattern consists of a random-dot texture which can be moved independently from the object. A white stationary screen (12° wide) is mounted between the stripe and the ground pattern. In all experiments represented in the figure, the oscillation amplitude of the stripe was $\pm 1^{\circ}$ (at 2.5 Hz frequency) and $\pm 0.5^{\circ}$ for the random texture (when oscillating). The standard responses (dotted horizontal line) measures the attractiveness of the stripe when oscillated alone, while the random texture was stationary. When object and ground are bouth oscillated with the same frequency, the average attraction towards the stripe depends on the relative phase, as shown in the left side of the figure. When object and ground are oscillated with different frequencies (2.5 and 1.8 Hz, respectively), the average attraction is about "standard". Each point is the mean of ten individual measurements. Each individual measurement lasted 2 min. The vertical bars denote standard errors of the mean. The continuous line is given by (hand) fitting of the experimental data with equation (3), where k_o is determined by the "standard" response ($k_o=1$) and k_4 is the free parameter. From Poggio and Reichardt (1976).

quency ω and relative phase ϕ. The average attractiveness of the object (defined by the time average of the fly's torque) is measured in units of the standard response to the stripe oscillating in front of the stationary ground texture. For equal frequencies of oscillation (the amplitude is 0.5° for the ground and 1° for the stripe). Fig. 6 shows that the detection of the stripe is reduced when the phase is either $\phi = 0$ (in phase, the two movements are "coherent") or $\phi = \pi$ (antiphase, the two movements are in phase opposition). The attractiveness of the object reaches its maximum for $\phi = \frac{1}{2}\pi$ and is also strong when object and ground oscillate with different frequencies. An opaque white screen interposed between the stripe and the ground, as shown in the inset of Fig. 6, does not have any influence on the effect if the screen is not too wide (not more than 20°). This indicates the existence of lateral (nonlinear) interactions between receptors stimulated by the object and receptors stimulated by the ground. Another version of this experiment is shown in Fig. 7. Whereas equal frequencies of oscillation were used, the amplitudes for the ground and for the object (stripe) were this time equal ($\pm 1^{\circ}$). As can be read from the phase dependence of the average responses, plotted in Fig. 7, the object is not detected for phases $\phi = 0$ and $\phi = \pi$ whereas the attractiveness of the object reaches its maximum at $\phi = \frac{1}{2}\pi$ and $\phi = \frac{3}{2}\pi$. It can be shown that for small osciallation amplitudes (as the ones used in the experiments plotted in Fig. 6 and 7) the light signals onto the receptors are periodic functions of time, containing mainly the first harmonic of the oscillation frequency. Thus, at least for a qualitative discussion, we may neglect higher harmonics. Under this assumption fourth-order interactions of the kind shown in Fig. 5 fully account for the average responses in Fig. 6 and Fig. 7, which are in fact rather well fitted by the typical fourth-order response

$$\bar{y} = k_o + k_4 \cos 2\phi \tag{3}$$

with $k_o > 0$ being the contribution from the excitatory self- or crossinteractions and $k_4 < 0$ the contribution from the

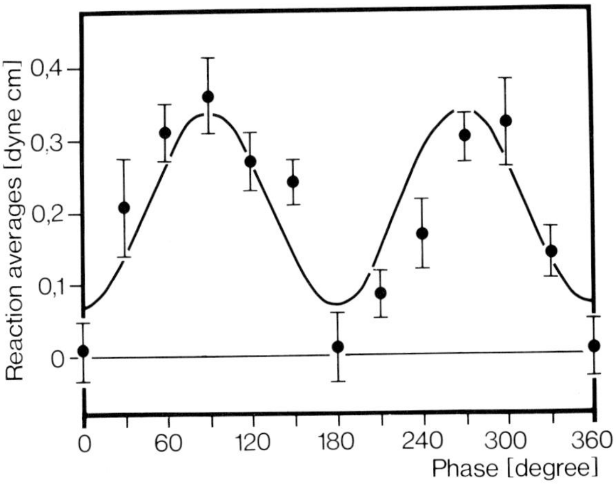

Fig. 7: Experimental conditions as described in the legend and the inset of Fig. 6, except for the equal oscillation amplitudes of $\pm 1^{o}$ for the stripe and the noise ground. Each point represents the average from 10 flies. The average responses are given in absolute units. The vertical bars denote standard errors of the mean. The continuous line is the component $k_4 \cos 2\phi$ in equation (3), derived from a Fourier-analysis of the data plotted in the figure.

fourth-order inhibitory crossinteraction; $k_o = -k_4$ for equal amplitudes of background and object oscillations. ϕ in equation (3) represents the phase difference between the sinusoidal motions of object and ground. Detailed calculations show (Poggio and Reichardt, in preparation) that the graphs in Fig. 5 are consistent with these experimental results. Moreover, the simplest interactive network which can discriminate relative motion of object and ground is the fourth-order one of Fig. 5a.

Let us now discuss the main steps of the argument: Self-interactions provide the "excitatory" attraction towards the stripe. What is additionally needed is a mean inhibitory influence, effective for "coherent" oscillations of the two patterns and ineffective for "incoherent" motion. The lowest degree of lateral interactions which should be considered is degree two; the next higher ones have degree four.

The only second-order interaction graph which needs to be considered is a graph with symmetry properties and two inputs, one stimulated by the stripe, the other by the ground texture. However, when the inputs receive two sinusoids with equal frequencies but different phases, the time averaged output is periodic in the phase-angle with a periodicity of 2π and not of π. In addition, the ensemble average of the mean outputs of nets consisting of these interactions is zero because of the random nature of the texture and because of the spatial distribution of the stripe first harmonic.

It can be shown, on the other hand, that the fourth-order interactions of Fig. 5 satisfy the above requirements. An oscillation of the stripe at frequency ω generates a periodic signal of double frequency at the output of the second-order self- (a) or cross- (b) interactions of Fig. 5. Oscillation of the noise yields a similar result. For simplicity, the case of the self-interactions (Fig. 5a) is considered. The important point here is that the frequency and phase of the ensemble average of the quadratic operations (self-interactions) depends on the frequency and phase of the motion and not on the structure of the pattern if the oscillation amplitude is small. Due to the frequency doubling, phase differences ϕ between the oscillations of the stripe and the texture are mapped into a double phase difference at the input of the second-order cross-interaction. Thus, the mean output of the latter interaction depends on 2ϕ. The additional assumption that this output is "inhibitory" ($k_4 < 0$) whereas the output of the self-interactions is "excitatory" ($k_o > 0$, attraction towards the object) yields equation (3).

Even in terms of the simplified analysis outlined here it is obvious that an increasing ground amplitude should increase the effect of the lateral inhibitions. Fig. 8 shows that this is indeed the case. For $\Phi = 0$ or $\phi = \pi$ the attraction of the object reduces to zero for equal amplitudes of stripe and

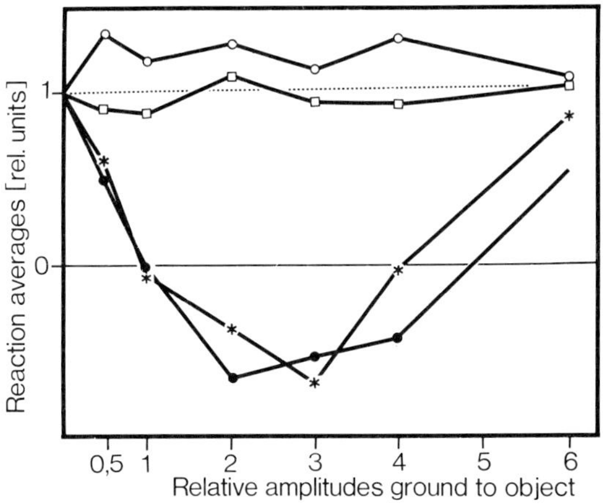

Fig. 8: Experimental details as described in the legend of Fig. 7, except for the various oscillation amplitudes of the ground with respect to the stripe's amplitude of $\pm 1^{\circ}$. The relative phase relations between stripe and ground are: 0° (●●●), 90° (ooo), 180° (∗∗∗). □□□ designate measurements with different frequencies (2.5 Hz for stripe and 1.8 Hz for random-dot texture). The average standard error of the means is ± 0.1 relative units. From Poggio and Reichardt (1976).

ground. For increasing ground amplitudes the inhibition overrides the excitatory contributions of the self-interactions. The fly is then repelled by the stripe. In an equivalent way, for $\phi = \frac{1}{2}\pi$ the inhibition turns into excitation and the attractivity of the stripe increases. At large amplitudes a number of factors reduce the effect. Higher harmonics now become significant and higher order terms could play a larger role.

So far I have discussed the computational properties of the visual system which enable the fly to discriminate moving objects from random-dot textured grounds. More recently the experiments were extended to figures consisting of random-dot contrastet vertical stripes (5° to 22° width) moved in front of a ground of the same random-dot texture. A typical example of

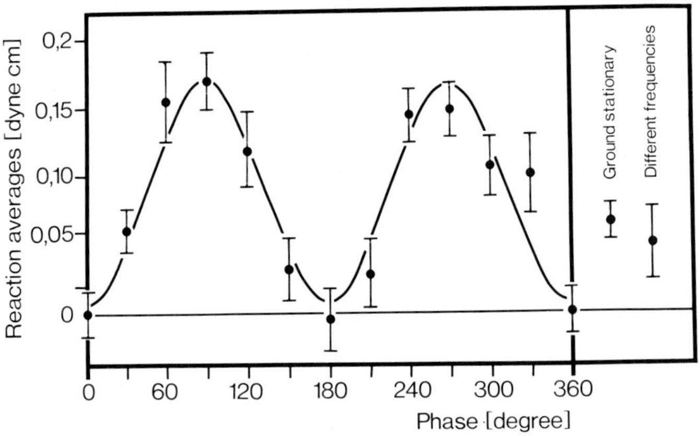

Fig. 9: The experimental conditions are in so far different from those described in the legends of Fig. 6 and Fig. 7 as the stimulus of figure and ground was a noise pattern displayed on two oscilloscope tubes. The figure consisted of a vertically oriented noise stripe, 10° wide. As before, figure and ground were moved horizontally, however, with equal amplitudes of $\pm 5^{\circ}$. The average position of the figure amounted to $\psi_o = \pm 30^{\circ}$.

The left side of the plot contains the phase dependence of the reactions, whereas the right side gives the responses to the oscillating figure and ground for different frequencies: 2.5 Hz for the figure and 1.8 Hz for the ground. Each point represents the average from 10 flies. The responses are given in absolute units. The vertical bars denote standard errors of the mean. The continuous line is the component $k_4 \cos 2\phi$ in equation (3), derived from a Fourier-analysis of the data plotted in the figure.

such a phase experiment is shown in Fig. 9. The width of the random-dot stripe amounted to 10°; the amplitudes for the stripe and for the ground were equal ($\pm 5^{\circ}$). Except for the average reaction amplitudes, the phase profile, shown in Fig. 9, does not basically differ from the object-ground profile reported in Fig. 7. This result suggests that the mechanisms responsible for object-ground as well as for figure-ground separation are the same.

Different oscillation frequencies of object and ground or figure and ground lead to a zero mean contribution from the fourth-order crossinhibitions. The contributions of the self-interactions, elicited by the ground, cancel, because (beside being homogeneously distributed on the two eyes) they are counteracted by those fourth-order inhibitory interactions whose inputs are all stimulated by the (in-phase) ground texture. Consequently, only the "excitatory" contributions elicited by the stripe or figure are left as in the case of a stationary background. The results from two sets of experiments with stationary ground and with two different oscillation frequencies for the figure and the ground are also plotte in Fig. 9. The reaction averages are not significantly different from each other.

These experimental findings and their analysis can be summarized by a simplified scheme, presented in Fig. 10. In this case the figure consists of a random-dot ring and the ground of the same random texture, so that if figure and ground are not moving against each other, the ring disappears in the ground. This impression does not change when figure and ground are moved together relative to the fly's eyes, since every excitation induced by a single contrast element is counteracted by an inhibition of another contrast element. However, as soon as the figure is moved relative to the ground (either with a different oscillation frequency, or in statistical independence from the motion of the ground), the mean inhibitory contributions across the boundaries of the figure drop to zero. Consequently, the part of the nerve net receiving optical stimulation from the figure becomes independent from the influence of the part which receives stimulations from the ground. That is to say, the object- or figure-ground discrimination effect consists in a context (ground) independent perception of the object or the figure.

The experimental results reported so far are in accordance with both interaction schemes of Fig. 5. The most elementary scheme presented in Fig. 5a contains two receptors whereas the

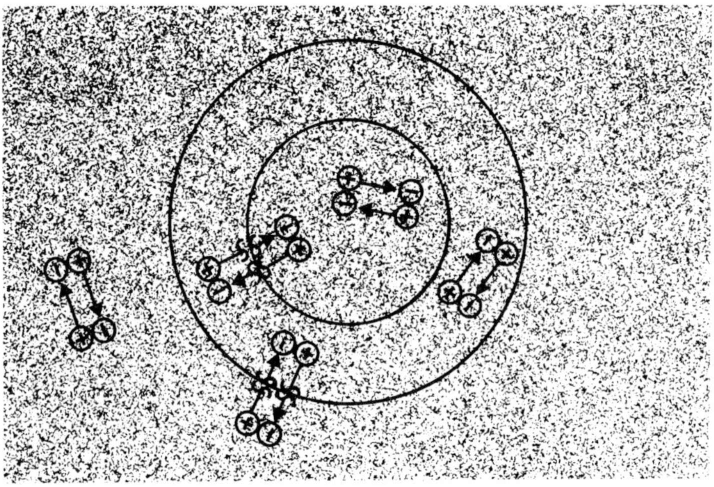

Fig. 10: A simplified version of the figure-ground discimination effect. Figure (ring) and ground are textured with the same noise contrast. Without relative motion between figure and ground, the figure disappears in the textured ground. However, when ring and ground are moved relative to each other, the boundaries of the ring are easily visible. When a fly is moving relative to the pattern shown here, each contrast element contributes with an attractiveness component, designated with (+). The signals received at the light-receptor level from different contrast elements are coherent if they originate from contrast elements which are not moving relative to each other. They induce inhibitory contributions (-) which compensate the excitatory (+) ones. The condition of coherency is fulfilled by the contrast elements of the ground as well as by the contrast elements located in the ring. However, coherency breaks down and consequently inhibition (-) is zeroed in the mean across the two boundaries of the ring when the figure (ring) is moved relative to the ground; this effect is indicated by the double wavy lines SS. The figure-ground discrimination therefore rests on the fact that the part of the nerve net which receives stimulations from the figure becomes independent from the influence of the part which receives stimulations from the ground. This is the explanation for the observation that the perception of the figure becomes context(ground) independent.

the scheme in Fig. 5b is driven by four receptor inputs. Consequently, the scheme of Fig. 5a would be insensitive to the

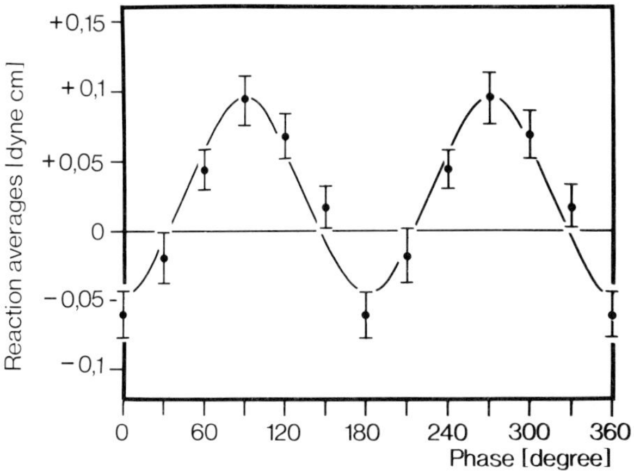

Fig. 11: All details as reported in the legend of Fig. 9, except that the figure (10° wide noise stripe) is oscillated vertically. Phase zero is defined such that if the ground moves to the right the figure moves upward.

orientation of motion of figure and ground. This situation, however, is quite different for the four receptor input scheme in Fig. 5b as the double receptor inputs should create an orientation sensitivity, except for the special case that many of such interaction schemes are homogeneously distributed over all possible orientation directions. In order to test whether the two or the four receptor model is realized in the fly, figure and ground had to be moved in different directions. Should the average responses of the phase profiles turn out to be different from the ones reported so far (ground and figure moved horizontally), the scheme presented in Fig. 5a had to be discarded. Fig. 11 shows the phase profile of the averaged responses to a horizontally oscillating ground and a vertically oscillating figure, both of the same noise texture. As already specified in connection with Fig. 9, the frequency of both oscillations amounted to 2.5 Hz and the oscillation amplitudes to $\pm 5^{o}$. The center of the figure was again located at the average position $\psi_o = \pm 30^{o}$.

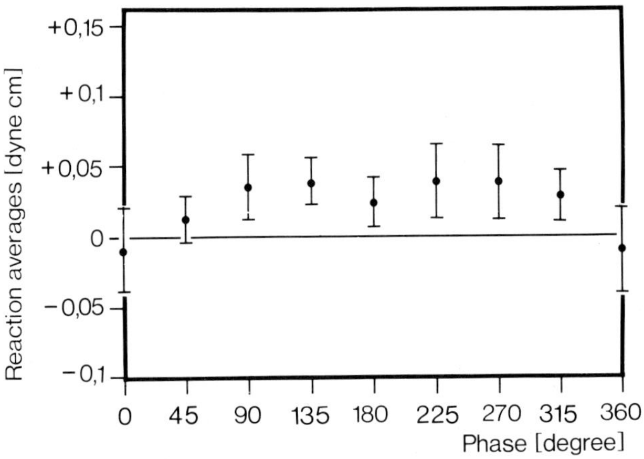

Fig. 12: All details as reported in the legend of Fig. 9, except that figure and ground were moved vertically.

In this experiment phase zero is defined by the following condition: if the ground moves to the right, the figure moves upward. The result of the experiment shows that the phase profile is again periodic in π, as reported before for horizontally moving figures and ground. The response, however, becomes negative (the fly is repelled) for phase angles 0, π and 2π, a result which contradicts the two receptor model. The changes observed in the phase profile are even more dramatic when the figure and the ground are both moved vertically. The results from such an experiment, which has been carried out with the same parameters, are shown in Fig. 12.

The experimental findings strongly suggest that the four receptors interaction scheme, presented in Fig. 5b is responsible for figure-ground discrimination by the fly. The main orientation dependent effect seems to come from the orientation of the ground, whereas the orientation of the figure seems to be less critical.

The next logical step in the analysis of the figure-

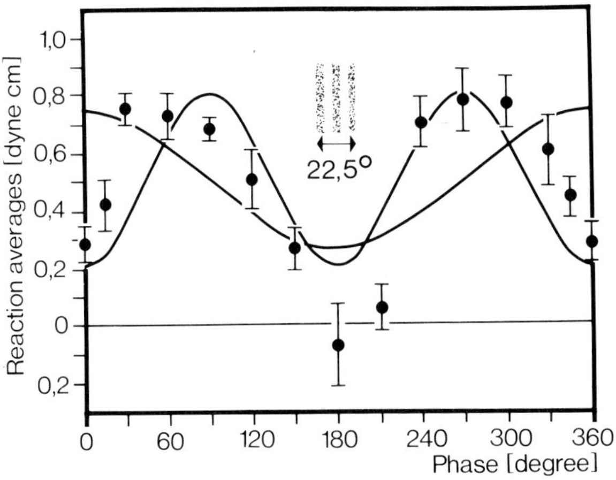

Fig. 13: In this experiment the figure consisted of a 22.5° wide pattern of alternating noise and white stripes, whereas the noise texture of the ground remained the same. Figure and ground were oscillated horizontally with 2.5 Hz and equal oscillation amplitudes of $\pm 3.5^{\circ}$. The average position of the figure amounted to $\psi_o = \pm 30^{\circ}$.

The plot shows the phase dependence of the average reactions. Each point represents the average from 10 flies. The responses are given in absolute units. The vertical bars denote standard errors of the mean. The continuous lines are the components $k_2 \cos \phi$ and $k_4 \cos 2\phi$ in equation (4), derived from a Fourier analysis of the data plotted in the figure.

ground discrimination problem was the application of figures which did not consist of the same contrast texture as the ground. Fig. 13 shows the outcome of such an experiment where the figure consisted of small alternating random-dot and white stripes moved horizontally in different phase relations with respect to the horizontally moving random-dot textured ground. The result shows that the flies are already attracted by the figure when figure and ground are moved in synchrony ($\phi = 0$). The attraction increases when the phase is either in the region near $\pi = \frac{1}{2}\pi$ or $\phi = \frac{3}{2}\pi$ and becomes slightly negative or zero for $\phi = \pi$. The

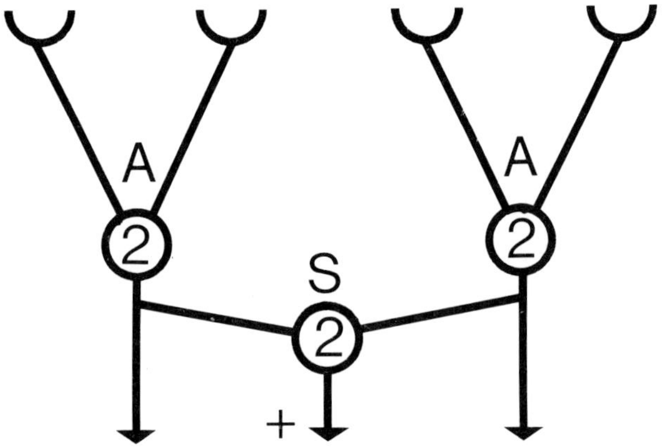

Fig. 14: Graphical representation of a possible interaction structure underlying the computation of the $k_2 \cos \phi$ component in equation (4). The results of two antisymmetric second-order crossinteractions of the type discussed in connection with Fig. 1a are crosscorrelated by a second-order symmetric crossinteraction whose contribution is excitatory.

important new information we gain from this experiment is that the dependence of the average reaction on the phase does not only contain a component with a periodicity of π, but in addition at least a second, even strong component with a periodicity of 2π so that instead of equation (3) the response presented in Fig. 13 can be approximated by

$$(4) \quad \bar{y} = k_o + k_2 \cos \phi + k_4 \cos 2\phi \quad \text{with } k_o>0; k_2>0 \text{ and } k_4<0$$

It can be easily shown that the term $k_2 \cos \phi$ can not result from a second-order symmetric interaction. The only possible candidate at present is a second-order symmetric cross-interaction operating between two second-order antisymmetric cross-interactions of the type as presented in Fig. 1. A graphical representation of this interaction is given in Fig. 14. The contribution of this interaction component is excitatory.

An important point for the analysis of the object-ground and figure-ground discrimination is the assumption that - like in the case of position computation - stationary images on the retina, irrespective of their brightness, have no influence on the response. If this is so, the computations involved in the discrimination processes would only operate on receptor outputs which result from changes of light fluxes at the receptor inputs. In order to test this assumption, a sequence of experiments (together with controls) was undertaken with stationary stripes (6° or 12° angular width) whose surfaces were either random-dot, black or white. The ground always consisted of a random-dot texture which was oscillated with a frequency of 2.5 Hz and an oscillation amplitude of $\pm 1.4^{o}$. The results of these experiments are plotted in Fig. 15. Let us first consider the three control experiments. The open circles represent reaction averages to in-phase oscillations of ground and random-dot stripes. These responses are zero. When only the random-dot stripe (either 6° or 12° wide) are oscillated with an amplitude of $\pm 1.4^{o}$ and the ground is at rest, the average response (filled circles) increases about linearly with increasing stripe width. The same results are obtained when figure and ground are oscillated with equal amplitudes, but different frequencies (2.5 Hz for the stripe and 1.8 Hz for the ground). As has been discussed before, under these conditions, the mutual inhibitory interactions do not contribute to the response and consequently the effect of the stripe (figure) becomes independent from the ground. This is the reason why the strength of the reaction (open squares) is the same as in the case when only the figure is oscillated and the ground is at rest. The three other reaction averages contain the essential information. In these experiments the stripe (figure) was at rest and only the ground was oscillated. The stars refer to the condition that the figure consisted of a random-dot texture, the open triangles to a white figure and the filled triangles to a black figure. The fact that these reaction averages are zero and independent of the texture and the brightness of the stripe tells us that only brightness changes which result from motions or from flicker

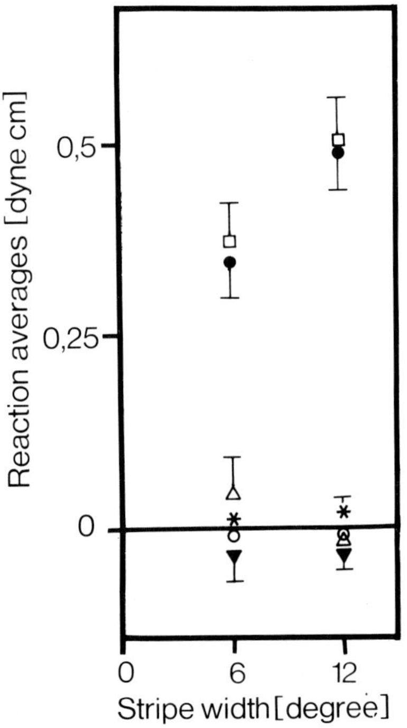

Fig. 15: Reaction averages to stripes (figures) and/or a noise textured ground. The stripes were positioned or when oscillated average positioned at $\psi_o = \pm 30^o$. The oscillation frequency amounted to 2.5 Hz and in the case of the two frequencies experiment 2.5 Hz and 1.8 Hz. Each point is the reaction average from 10 flies. The vertical bars denote the standard errors of the mean. The different points refer to the following experimental conditions (for details see text):

o Control
● Noise stripe oscillated, noise ground fixed
□ Noise stripe and noise ground oscillated with two different frequencies
▼ Black stripe stationary, noise ground oscillated
Δ White stripe stationary, noise ground oscillated
* Noise stripe stationary, noise ground oscillated.

lead to functional components which influence the object-ground and figure-ground computations.

Discussion

The experiments reported in this paper lead to the conclusion that fourth-order lateral interactions, involved in the orientation response of the fly, can perform nontrivial computations which at least extend to the level of figure-ground discriminations. They almost certainly may also play an important role in the fly's ability to discriminate and to perceive patterns.

It should be mentioned here that the experiments were not only restricted to the front parts of the two compound eyes but also extended to the side regions of the eyes. The responses measured in these regions are not different from those reported here, except that the amplitudes of the averaged reactions are smaller than those recorded from the front regions.

A closed-loop discrimination of an object in front of a ground texture (Virsik and Reichardt, 1976), when "incoherent" relative motion takes place, can be explained in terms of the computations which account for the open-loop object-ground effect. The parameter, which critically determines the extent of object masking by the ground texture, is the coherence between light modulations due to the movements of the object and the ground. Therefore, object discrimination depends on the kind of relative motion between the two patterns and of course on the structure of the pattern.

Our investigations show the intrinsic spatio-temporal nature of movement and position computations in the fly. Position information is essentially obtained from the location of the stimulus on the retina. The behavioural response (torque) is spatially coded through the "weight" $D^*(\psi)$, see Fig. 2. The temporal nature of the input is, however, also important: a stabilized retinal image does not elicit any mean attraction in the fly. The critical interplay of spatial and temporal parameters is typically shown in the object-ground and figure-ground experiments. Either the amplitude (a spatial parameter) or the phase (a temporal parameter) of object (figure) and ground oscillations can turn attraction into repulsion. This shows the inadequacy of simple geometrical parameters to characterize spontaneous pattern preference in insects as well as in other animals. The spatio-temporal mapping of a stimulus onto the receptor array plays an imporant role in the lateral inhibitory network of Limulus (see Hartline and Ratliff, 1972). The essential new point which emerges from this work is the computational role of nonlinear interactions in a spatially distributed processing system. The information processing capabilities of nonlinear lateral inhibitions can be essentially greater than in the quasi-linear case of Limulus. While the response of a linear system to a complex input is the mere superposition of the responses to elementary components, the behaviour of a nonlinear system can be very different from a simple superposition of elementary effects. The object-ground and the figure-ground discrimination described here is an illustration of this property. There is absolutely no linear filter process which could produce the same effect.

A spatial pattern may be described in terms of a complete set of basic elements like spatial frequencies or line segments. If the pattern processing system obeys spatial superposition each such element may be correctly considered a "feature" since the output of the system is completely known from knowledge of the response to the individual elements. Lateral nonlinear interactions drastically change this picture. The spatial extent and organization of the integrations define complex "features" composed of pattern elements which become interrelated by the interactions. Only beyond the nonlinear interaction range are two complex features again independent elements of a pattern. In other words, lateral interactions define the "features" in which a pattern can be decomposed. The elementary system for object- and figure-ground discrimination consists of nonlinear, lateral interactions which receive their inputs from each of two receptors whose orientations are different. However, a "feature" property, determined by these interactions, can be dramatically changed if an object or figure is moved relative to the ground. Under this operation the contribution of the lateral interactions may become zero (in the time average) across the boundary of the figure, so that the figure behaves as a context independent "feature".

The localization of the interactions, discussed in this paper, is still an unsolved problem. The movement computation is associated to antisymmetric, second-order interactions. The position or attractiveness computation rests on two types of mechanisms, local nonlinear flicker detectors (selfinteractions, see Fig. 1b) and lateral, fourth-order symmetric interactions (see graphs in Fig. 5a and Fig. 5b). Distinct functional and computational properties are associated to these four types of interactions. The functional separation of four types of interactions does not necessarily imply four distinct neural circuitries. However, some arguments suggest that distinct mechanisms may indeed exist. While the movement interaction seems actually separated from the attractiveness interactions, a similar structural separation between the self- and fourth-order crossinteractions still appears doubtful.

Strictly connected to the problem of a structural separation of the interaction types, localization of the underlying mechanisms in the fly's brain is still an open question. The small delay involved in the fly's behavioural reaction (20 msec) suggest that possibly, but not necessarily, all four interactions require a small number of sequential steps. Antisymmetric, second-order interactions probably take place at the level of the second optical ganglion, the medulla, where small field direction selective neurons have been recorded from. Spatial summation of these local detectors is probably performed in a part of the third optical ganglion, the lobula plate. The localization of the direct nonlinear self-interactions, which compute position information, is unknown. However, at least part of the processing is possibly performed in the medulla. Although neurophysiologists have tried to characterize movement detection in the fly, a similar attempt concerning position computation is still missing. Some electrophysiological data may be indirectly connected with the fourth-order interactions which underly the object- and figure-ground discrimination

(Collett, 1971; Palka, 1972; O'Shea and Rowell, 1975). In which area of the fly's brain these functional interactions are physiologically localized presents a challenging question.

Acknowledgments

I would like to thank my coworkers Dr. C. Wehrhahn and Dr. T. Poggio for constructive criticism, Mr. Heimburger for drawing the figures and Mrs. I. Geiss for typing the manuscript.

References

Buchner, E. (1974). Bewegungsperzeption in einem visuellen System mit gerastertem Eingang. Dissertation Eberhard-Karls-Universität Tübingen.

Buchner, E. (1976). Elementary movement detectors in an insect visual system. Biol. Cybernetics 24, 85-101.

Collett, T.S. (1971). Visual neurons for tracking moving targets. Nature, 232, 127-130.

Eckert, H.E. (1973). Optomotorische Untersuchungen am visuellen System der Stubenfliege *Musca domsetica* L. Kybernetik 14, 1-23.

Fermi, G. and Reichardt, W. (1963). Optomotorische Reaktionen der Fliege *Musca domestica*. Kybernetik 2, 15-28.

Geiger, G. and Poggio, T. (1975). The orientation of flies towards visual patterns: on the search for the underlying functional interactions. Biol. Cybernetics 17, 1-16

Götz, K.G. (1964). Optomotorische Untersuchungen des visuellen Systems einiger Augenmutanten der Fruchtfliege *Drosophila*. Kybernetik 2, 77-92.

Götz, K.G. (1972). Principles of optomotor reactions in insects. Biblthca ophthal. 82, 251-259.

Götz, K.G. (1975). The optomotor equilibrium of the *Drosophila* navigation system. J. comp. Physiol. 99, 187-210.

Hartline, H.K. and Ratliff, F. (1972). Inhibitory interaction in the retina of *Limulus*. In: Handbook of Sensory Physiology Vol. VII/2 (ed. M.G.F. Fuortes), pp. 381-448. Berlin, Heidelberg, New York: Springer-Verlag.

Hassenstein, B. (1951). Ommatidianraster und afferente Bewegungsintegration (Versuche am Rüsselkäfer *Chlorophanus viridis*). Z. vergl. Physiol. 33, 301-326.

Hassenstein, B. (1958). Über die Wahrnehmung der Bewegung von Figuren und unregelmässigen Helligkeitsmustern. Z. vergl. Physiol. 40, 556-592.

Hassenstein, B. (1959). Optokinetische Wirksamkeit bewegter periodischer Muster. Z. Naturf. 14b, 659-674.

Hassenstein, B. and Reichardt, W. (1956). Systemtheoretische Analyse der Zeit-, Reihenfolgen- und Vorzeichenauswertung bei der Bewegungsperzeption des Rüsselkäfers Chlorophanus. Z. Naturf. 11b, 513-524.

Heimburger, L., Poggio, T. and Reichardt, W. (1976). A special class of nonlinear interactions in the visual system of the fly. Biol. Cybernetics 21, 103-105.

Hengstenberg, R. and Götz, K.G. (1967). Der Einfluss des Schirmpigmentgehalts auf die Helligkeits- und Kontrastwahrnehmung von Drosophila-Augenmutanten. Kybernetik 3, 276-285.

Kirschfeld, K. (1972). The visual system of Musca: studies on optics, structure and function. In: Information Processing in the Visual System of Arthropods (ed. R. Wehner), pp. 61-74. Berlin, Heidelberg, New York: Springer-Verlag.

Marmarelis, P. and McCann, G.D. (1973). Development and application of white-noise modeling techniques for studies of insect visual nervous system. Kybernetik 12, 74-90.

McCann, G.D. and MacGinitie, G.F. (1965). Optomotor response studies of insect vision. Proc. R. Soc. B. 163, 369-401.

O'Shea, M. and Fraser Rowell, C.H. (1975). Protection from habituation by lateral inhibition. Nature 254, 53-54.

Palka, J. (1972). Moving movement detectors. Am. Zool. 12, 497-505.

Pick, B. (1974a). Visual flicker induces orientation behavior in the fly Musca. Z. Naturf. 29c, 310-312.

Pick, B. (1974b). Das stationäre Orientierungsverhalten der Fliege Musca. Dissertation Eberhard-Karls-Universität Tübingen.

Pick, B. (1976). Visual pattern discrimination as an element of the fly's orientation behaviour. Biol. Cybernetics 23, 171-180.

Pick, B. and Buchner, E. (1978). Movement-specific wide-angle interactions in the visual system of the fly. Biol. Cybernetics (to be submitted).

Poggio, T. (1974). Processing of visual information in flies: from a phenomenological model towards the nervous mechanisms. In: Atti della prima riunione Scientifica (Camogli, 1973). Soc. Ital. Biofis. Pura e Applicata, pp. 217-225.

Poggio, T. (1975). Processing of visual information in insects: outline of a theoretical characterization. In: Biokybernetik. Band V (Ed. H. Drischel and P. Dettmar), pp. 235-243. Jena: VEB Gustav Fischer Verlag.

Poggio, T. and Reichardt, W. (1973a). Considerations on models of movement detection. Kybernetik 13, 223-227.

Poggio, T. and Reichardt, W. (1973b). A theory of the pattern induced flight orientation of the fly Musca domestica. Kybernetik 12, 185-203.

Poggio, T. and Reichardt, W. (1976). Visual control of orientation behaviour in the fly. Part II. Towards the underlying neural interactions. Quart. Rev. Biophysics 9, 3, 377-438.

Poggio, T., and Reichardt, W. in preparation.

Reichardt, W. (1957). Autokorrelations-Auswertung als Funktionsprinzip des Zentralnervensystems. Z. Naturf. 12b, 448-457.
Reichardt, W. (1961). Autocorrelation; a principle for the evaluation of sensory information by the central nervous system. In: Sensory Communication (ed. W.A. Rosenblith), pp. 303-318. New York: John Wiley.
Reichardt, W. (1969). Movement perception in insects. In: Processing of Optical Data by Organisms and Machines (ed. W. Reichardt), pp. 465-493. London, New York: Academic Press.
Reichardt, W. (1970). The insect eye as a model for analysis of uptake, transduction, and processing of optical data in the nervous system. In: The Neurosciences, Second Study Program (ed. F.O. Schmitt), pp. 494-511. New York: Rockefeller University Press.
Reichardt, W. (1973). Musterinduzierte Flugorientierung. Verhaltens-Versuche an der Fliege Musca domestica. Naturwissenschaften 60, 122-138.
Reichardt, W. and Poggio, T. (1975). A theory of pattern induced flight orientation of the fly Musca domestica: II. Biol. Cybernetics 18, 69-80.
Reichardt, W. and Poggio, T. (1976). Visual control of orientation behaviour in the fly. Part I. A quantitative analysis. Quart. Rev. Biophysics 9, 3, 311-375.
Reichardt, W. and Varjú, D. (1959). Übertragungseigenschaften im Auswertesystem für das Bewegungssehen. Z. Naturf. 14b, 674-689.
Thorson, J. (1966a,b). Small signal analysis of a visual reflex in the locust: I, II. Kybernetik 3, 41-66.
Varjú, D. (1959). Optomotorische Reaktionen auf die Bewegung periodischer Helligkeitsmuster. Z. Naturf. 14b, 724-735.
Varjú, D. and Reichardt, W. (1967). Übertragungseigenschaften im Auswertesystem für das Bewegungssehen II. Z. Naturf. 22b, 1343-1351.
Virsik, R. and Reichardt, W. (1976). Detection and tracking of moving objects by the fly Musca domestica. Biol. Cybernetics 23, 83-98.
Wehrhahn, C. (1974). Verhaltensstudie zur musterorientierten Höhenorientierung der Fliege Musca domestica. Dissertation Eberhard-Karls-Universität Tübingen.
Wehrhahn, C. and Reichardt, W. (1975). Visually induced height orientation of the fly Musca domestica. Biol. Cybernetics 20, 37-50.

EVOLUTION OF THE ADAPTIVE LANDSCAPE*

Michael Conrad
Department of Computer and Communication Sciences
University of Michigan
Ann Arbor, Michigan

1. Introduction

Evolution is sometimes pictured in terms of hill climbing on an adaptive landscape (Wright, 1932), therefore as an optimization process. This imagery certainly should be treated with certain cautions--for one never knows the structure of the landscape, there is difficulty saying in general what precisely is to be optimized (i.e. it is difficult to define fitness), and it is possible that optima, even accessible optima, are never reached. However, there can hardly be any doubt that the mechanism of evolution through variation and natural selection potentially subserves an optimization process in some very general sense, that even if the products of evolution are not actually optimal they are qualitatively the most sophisticated forms in nature, and that the imagery of the adaptive landscape has been scientifically fruitful.

These considerations have suggested to a number of students of optimization that optimization algorithms based on mechanisms analogous to natural evolution might be particularly effective for hill climbing. Such algorithms have been developed most notably by Bremermann (1962) in the context of random search and selection and by Holland (1975) in terms of a general adaptive systems framework which can be specialized to include various types of genetic processes known to be important in evolution. There have also been a number of simulation studies of evolution per se

*This paper is dedicated to the memory of Ernst Pfaffelhuber, both in tribute to his contributions to physics, information theory, and biology and in rememberance of many stimulating conversations and perceptive suggestions.

(Barricelli, 1962; Reed et al., 1967), including studies which attempt to duplicate features of evolution without assuming a definite optimization problem or preassigning criteria for fitness (Conrad, 1969; Conrad and Pattee, 1970). The evolutionary approach to optimization is successful for certain classes of functions and now of practical importance (e.g. Bremermann, 1967a; Rechenberg, 1973). However, it cannot be said that evolutionary methods, or indeed any other known methods, are capable of handling problems outside a fairly restricted framework (in particular, not too high dimensionality, not too many local maxima or even no local maxima other than the global maximum). Likewise, simulations of evolution, while providing important biological insights, cannot be regarded as producing results in any way of comparable interest to those of natural evolution and in some cases (e.g. attempts to develop programs through evolution) are stubbornly unsuccessful (cf. Friedberg, 1958).

There are two possibilities. The first is that the evolutionary methods of optimization are not good methods, implying that we really basically misunderstand the process of organic evolution. There is (aside from the clear biological arguments against this point of view) a potentially critical mathematical argument. This is that in the absence of any restriction the class of optimization problems necessarily includes search problems equivalent in difficulty to a large class of problems (the so-called NP-complete problems) not one of which is known to be amenable to solution in an acceptable amount of time by a deterministic process (in the sense that the number of steps required for any known method of solution grows exponentially with the size of the problem, cf. Aho et al., 1974). While this does not necessarily imply that evolution methods are the best methods, either in general or for any restricted class of problems, it does imply that arguments against Darwinian evolution on the grounds of weakness of evolution methods apply at least as well to any alternative mechanism of adaptive evolution. Mathematical limits of information processing of the above type (including information processing concomitant to optimization) are objects of current active research (cf. Karp, 1972). Real physical limits, less well researched (but cf. Bremermann, 1962, 1967b; Landauer, 1976) are much more restrictive than these mathematical limits, so that when these are taken into account it must be admitted that the argument becomes still stronger.

The second possibility is that the theory of evolution is fundamentally correct but that the optimization problem in nature (despite the complexity of living systems) satisfies the conditions which make evolution methods effective. In this paper we show that this is in fact the true situation; in other words, that the adaptive landscape is just the type of landscape in which many of the hills and mountains are amenable to hill and mountain climbing through variation and natural selection. The argument has two parts. First we show that the condition for effective evolutionary hill climbing is that the ascent never requires more than a few simultaneous genetic events and that in general this means stepwise changes at the genetic level must in many cases be concomitant to slight variations of the phenotypes. In the second part we show (on biophysical grounds) that proteins can always undergo stepwise evolution to fulfill this condition. In this sense we can think of the adaptive landscape as fixed, but as being so organized (because of the physical properties of proteins) that it is always possible to evolve stepwise along one smooth dimension to a point in which smoothness of ascent in other dimensions is optimized, in which case we have evolution to regions of the adaptive landscape which are more amenable to evolutionary search.

2. Structure and Topology of the Adaptive Landscape

The adaptive landscape is picturesque terminology for a plot of fitness against environmental and biological variables, where the latter might be taken as phenotypic traits, but more usually are taken as the gene structures which code for these traits. Fitness is difficult to define precisely and is possibly best regarded as a primitive concept of biology (Conrad, 1972). However, it is frequently (and for the present purposes adequately) thought of in terms of the probability of contributing genes to future generations (cf. Waddington, 1968).

Most useful for this paper is a genotypic landscape which is more detailed than the classical gene frequency picture (which can be regarded as a special case). This fine structured landscape is a multidimensional space with one axis measuring fitness and the other axes describing base sequence in DNA. Four axes are assigned to the first position, one for each type of base, and likewise for each succeeding position.

The presence of a base is represented by a one on the axis and its absence by a zero, with a one on any of the four axes excluding a one on any of the others and zeros on all the axes indicating the end of the string (and the number of the genetic axis is always taken large enough to describe the longest physically and biologically realistic genome). Axes may also be added for environmental variables (global landscape) or these variables may be assumed fixed (local landscape). Clearly the fitness surface is not completely connected because of the discreteness of the possible values on the gene axis. An alternative, nondiscrete description is also possible, plotting fitness of populations against frequencies of bases in each position. The classical adaptive landscape is the special case in which the frequencies are frequencies of genes and the environmental axes are eliminated, thereby assigning fitnesses to different possible combinations of gene frequencies under the assumption of a constant but unspecified global environment. However, these less detailed descriptions do not show the potentially significant differences in the fitness associated with different base sequences or gene combinations compatible with the same frequencies. For a single structural gene the adaptive landscape can be redescribed in terms of primary structure of proteins, with twenty axes for each position (one for each amino acid) and one third as many filled positions. In the case of a single gene or protein the fitness axis should be regarded as a measure of the contribution of this gene to the fitness of the organism or population, all other genes being constant. This picture is simple to work with and will be useful.

Define the genetic distance between any two genes (or proteins) to be the Hamming distance between them (i.e. the minimal number of changes of zeros and ones required to make them identical). The degree of gradualism of any neighborhood of the fitness surface is the minimum genetic distance which any genome specified by projecting this surface on the genetic axes must traverse to reach a higher level of fitness. If for two such surfaces this minimum distance is both one, the surface for which fitness rises more slowly is equally gradual, but less steep. Sometimes we will say that pathways in the landscape with a higher degree of gradualism are more smoothed out, though technically a discrete, one step change is not a smooth change. Also, it is possible to define a second concept of gradualism, viz. involving single step changes which affect a number of genetic axes (e.g. block of nucleotides, entire genes, or block of genes) at once.

3. Optimal Structure of the Landscape

Well known is the historic debate between saltationist and gradualistic schools of evolutionary thought. While it is possible that catastrophic but functionally significant morphological changes have occurred with gradual change in genetic structure (cf. Thom, 1970), it would seem that the weight of both biological and mathematical evidence strongly disfavors any significant role for catastrophic genetic events. This is a central point of the so-called modern synthesis (cf. Simpson, 1949; Mayr, 1963) and is also supported by studies with evolutionary algorithms, for which it is known both in theory and in practice (cf. Bremermann, 1962) that the optimal mutation rate for evolutionary hill climbing is one mutation per generation (which can be broadly translated to one independent genetic event per organism per generation), which means that the adaptive landscape must be so structured that it can be ascended in this stepwise fashion. The intuitive consideration is obvious: the occurrence of one event has probability p, two events probability p^2, and so forth, which means that the simultaneous occurrence of any significant number of events has negligible probability.

Here what is important is to investigate how differences in the rate of evolution depend on the number of simultaneous genetic events required to make each step up an adaptive peak. This is tantamount to determining how steep the climb is (in terms of increasing fitness) to regions of the landscape more amenable to the evolution process. Treated generally, this would be a rather complex problem and would require a number of assumptions to make the answer concrete in any case. We are interested only in an in principle proof and therefore discuss the simplest possible situation, showing subsequently that any injection of greater realism into this in principle situation would only strengthen the argument.

Consider a simple gene G_u coding for a given enzyme (or protein species). G_u is a sequence of bases $b_1,\ldots,b_{(3n)}$, where each b_i is one of four types of nucleotides in DNA. G_u codes for the sequence of amino acids (or primary structure) of the protein. Thus we can write $G_u \rightarrow S_u$, where $S_u = a_1,\ldots,a_n$ and each a is one of the approximately twenty types of amino acids in protein. For simplicity we work with S_u directly rather than with G_u.

Suppose that G_o has fitness W_o, or more precisely, that the genome which carries a gene coding for S_o has fitness W_o. Suppose also that to make minimal progress to the next higher point on the adaptive landscape (i.e. to go from fitness W_o to the next higher fitness level W_m) requires m alterations in the amino acid sequence of S_o, i.e. the genetic distance between G_o and G_m is between m and 3m, with approximately 3^n of the possible $4^{(3n)}$ possible sequences coding for S_m. The only way to jump between the two maxima is to do so by m simultaneous changes in the appropriate amino acids (along with no changes in any other amino acids). The probability for such an occurrence is

$$p(S_o \longrightarrow S_m) = \frac{N_1}{N_o} = \frac{p^m(1-p)^{n-m}}{19^m} \qquad (1)$$

where N_o is the number of organisms with gene G_o at time t, N_1 is the expected number of organisms with gene G_m after one generation, and p is the probability of a single amino acid change (e.g. the mutation probability). The probability of an appropriate amino acid change, however is p/19 since the change could be to any of the other nineteen amino acids. The average number of generations required for the appearance of protein S_m is thus

$$\bar{\tau}_{om}^{(m)} = \frac{1}{N_o p(S_o^m \xrightarrow{m} S_m)} = \frac{19^m}{N_o p^m (1-p)^{n-m}} \qquad (2)$$

where (m) indicate the number of required simultaneous genetic events and $\tau_{om} < 1$ would mean that more than one S_m is expected to appear after one generation (i.e. $N_1 > 1$). Note that p/19 is actually an average probability because of the redundancy structure of the code, i.e. not all transitions between amino acids are precisely equally probable. Also note that we make the reasonable assumption that the probabilities are independent. In fact any other assumption would increase the number of expected generations required for the appearance of any arbitrarily chosen sequence.

Now suppose that of the m! possible ways in which S_o can change into S_m by m single changes in amino acid sequence, there is one for which each W_i in the sequence is at least slightly greater than its predecessor, i.e. that

$$S_o \xrightarrow{1} S_1 \xrightarrow{1} \cdots \xrightarrow{1} S_i \xrightarrow{1} S_{i(i+1)} \xrightarrow{1} \cdots \xrightarrow{1} S_{m-1} \xrightarrow{1} S_m \qquad (3)$$

such that $W_i < W_{i+1}$ for all i= 0 to m. The probability for the appearance of S_m now depends on the probability of m appropriate single step mutations. The expected number of generations required for any single step in this chain is just a special case of Eq. (2),

$$\bar{\tau}^{(1)}_{i(i+1)} = \frac{19}{N_i(t)p(1-p)^{n-1}}, \qquad (4)$$

except that N_i is now a function of time since a new population (e.g. population i carrying gene G_i) begins to grow as soon as its appears. The average number of generations required for the appearance of S_m is now the sum of all the individual evolution times:

$$\bar{\tau}^{(1)}_{om} = \sum_{i=1}^{m} \bar{\tau}^{(1)}_{i(i+1)} = \sum_{i=1}^{m} \frac{19}{N_i(t)p(1-p)^{n-1}} \qquad (5)$$

To determine the expected evolution time it is clearly necessary to know the $N_i(t)$, i.e. how fast each population in the series grows. This is complicated, however, since the population size changes in a way which depends on the increase in fitness following on the appearance of S_i, on the appearance of alternate alleles or inevitable changes at other loci (both ignored). To keep things simple, we again make the most unfavorable possible assumption, viz. that mutation (and all other mechanisms of genetic change) are turned off until the new population effectively grows to the same size as the old population (i.e. until N_i grows to N_o), and that there is no further growth. Then Eq. (5) becomes

$$\bar{\tau}^{(1)}_{om} = \frac{19m}{N_o p(1-p)^{n-1}} + (m-1)D, \quad m \geq 1 \qquad (6)$$

where D is the delay time (or number of generations) before we allow mutation to be turned on and it is not necessary to consider delay following the first appearance of S_m. In general delay would be different for each step in the series, but it can always be taken sufficiently large to ensure that $\bar{\tau}_{om}^{(1)}$ is an underestimate (aside from the underestimation inherent in assuming no mutation until N_i reaches N_o).

Relative evolution times for the stepwise and simultaneous occurrence of m changes are given by the ratio

$$F(m) = \frac{\bar{\tau}_{om}^{(1)}}{\bar{\tau}_{om}^{(m)}} = p^{m-1} \left[\frac{19m + (m-1)D\, N_o p(1-p)^{n-1}}{19^m(1-p)^{m-1}}\right] \qquad (7)$$

A reasonable choice for mutation rate is $p = 10^{-8}$. Population size and differential growth rates (expressed in the values of D) could vary widely. A range of values of the ratio taken over a range of the population sizes from 10 to 10^9, a range of rates of D from 1 to 10^7, a range of values of n from 10 to 1,100, and of m from 1 to 8. (for $p = 10^{-6}, 10^{-8}$, and 10^{-10}) are given in Tables I-IV. The conclusion is quite robust and hardly surprising that the stepwise mode of evolution is astronomically faster than the catastrophic, simultaneous mode. For any reasonable choice of generation time a simultaneous change of, for example, five steps would take longer than the known age of the earth.

In the Appendix it is shown that the above general behavior is valid for all realistic values of p, and indeed for unrealistically large values of p. Tables I and II also show that F(m) decreases so much faster with increasing m than with increasing n that the relative advantage of the stepwise over the simultaneous mode of evolution decreases only minutely even if the former entails longer sequences and more genetic steps. More interesting, the difference between 1/F(m) and 1/F(m+1) is approximately equal to 1/F(m) and therefore increases with dramatic sharpness as m increases. This is also discussed in the Appendix, but can be seen from the exponential dependence of F(m) on m in Table I. This is important, for it means that the selective advantage of an increase in the degree of gradualism increases as the degree of gradualism departs from the one step case. In short, the one step strategy

m \ p	10^{-6}	10^{-8}	10^{-10}
1	1.0	1.0	1.0
2	2.9×10^{-6}	1.3×10^{-9}	1.1×10^{-11}
3	3.0×10^{-13}	1.1×10^{-18}	8.3×10^{-23}
4	2.4×10^{-20}	8.1×10^{-28}	5.9×10^{-34}
5	1.7×10^{-27}	5.5×10^{-37}	3.9×10^{-45}
6	1.1×10^{-34}	3.5×10^{-46}	2.4×10^{-56}
7	6.9×10^{-42}	2.2×10^{-55}	1.5×10^{-67}
8	4.2×10^{-49}	1.3×10^{-64}	9.0×10^{-79}

Table I. Variation of F(m) with step length and mutation rate. D=1000, $w=10^6$, n=300.

n	F(2)
10	1.3296399×10^{-9}
100	1.3296396×10^{-9}
300	1.3296391×10^{-9}
1100	1.3296369×10^{-9}

Table II. Insensitivity of variation of F(m) with protein length for m=2. $p=10^8$, D=1000, $N=10^6$.

N	F(2)
10	1.05623×10^{-9}
10^2	1.05266×10^{-9}
10^3	1.05291×10^{-9}
10^4	1.05540×10^{-9}
10^5	1.08033×10^{-9}
10^6	1.32964×10^{-9}
10^7	3.82271×10^{-9}
10^8	2.87534×10^{-8}
10^9	2.78060×10^{-7}

Table III. Variation of F(m) with population size for m=2. p=10-8, D=1000, n=300.

D	F(2)	F(3)
1	1.1×10^{-9}	8.3×10^{-19}
10	1.1×10^{-9}	8.3×10^{-19}
10^2	1.1×10^{-9}	8.6×10^{-19}
10^3	1.3×10^{-9}	1.1×10^{-18}
10^4	3.8×10^{-9}	3.7×10^{-18}
10^5	2.9×10^{-8}	3.0×10^{-17}
10^6	2.8×10^{-7}	2.9×10^{-16}
10^7	2.8×10^{-6}	2.9×10^{-15}

Table IV. Variation of F(m) with delay time for m=2 and 3. p=10-8, N=6, n=300.

is not only optimal, but a global attractor in the sense that the evolutionary advantage (in terms of expected evolution times) for approaching it is strong when close and even stronger when less close.

4. Optimizability of the Landscape

The basic conclusions of the previous section are:

(i) The potential rate of evolution is a maximum when the evolution is capable of proceeding step by step and decreases rapidly as the number of required simultaneous events increases.

(ii) The increase in relative evolution time for pathways of evolution requiring simultaneous and stepwise genetic events increases as step length increases, implying that the difference in the actual evolution times increases more rather than less rapidly. This means that selection for decrease will become more intense as the step length departs from one (all other things constant).

(iii) The above conclusions hold even if the total number of required changes increases with decrease in step length, or if the length of the primary sequence increases.

The important question is whether biological systems can evolve to fulfill these optimal conditions, i.e. whether the topology of the adaptive landscape is in some sense implicit in their organization and therefore itself amenable to evolutionary optimization. This is not a mathematical question, but rather a matter of the physical and organizational properties of biological matter. The basic consideration is that the functional significance of each primary structure (S_i) is determined by the physical-chemical properties, in particular the three-dimensional shape and charge distribution, implicit in this structure. The justification for this statement derives from the uniqueness of the folding process, and from the importance of certain features of molecular geometry for the molecular pattern recognition associated with specific catalysis and self-assembly (cf. Perutz, 1962). The folding process itself is fundamentally a matter of free energy minimization, i.e. the final conformation is presumably the conformation which minimizes energy and maximizes the number of possible configurational microstates, or at least is the most accessible such conformation given any reasonable initial conformation of the molecule or perturbed conformation from which renaturation is possible.

These facts, well known in themselves, have important implications for evolution. Since the properties of the protein (e.g. three dimensional shape and therefore function) are implicit in primary structure, it is in principle possible for the gradualism with which these properties change with slight change in primary structure also to be implicit in this structure--in which case this gradualism automatically becomes a trait susceptible to selective evolution. It is not difficult to characterize the forms of molecular organization which would be capable of supporting such gradualism. The key trait is the free energy of folding implicit in the primary structure. If this free energy of folding is increased (by selection) in such a way that features of shape important for recognition, catalysis, and control are overdetermined, these features will be more gracefully altered by typical (but noncritical) changes in primary sequence. Such overdetermination is mediated by redundancy in the various forms of weak bonding responsible for folding, e.g. van der Waal's interactions, disulphide bonds, or by increased utilization of amino acids with an increased number of structurally similar analogs. It may be combined with critical hinges (e.g. anomalous amino acids such as proline) in the primary structure, in which case certain definite three dimensional relationships are adjustable in a systematic way. The classic and extreme example are the immunoglobins (cf. Roitt, 1974). However, enzymes not required to exhibit numerous subtle gradations in the course of ontogenetic development also seem to be designed in such a way that gradualism is possible. One example are the dehydrogenases, known to exhibit greater evolutionary conservativism at the level of tertiary structure than at the level of primary structure (Rossman et al., 1975). Changes involving between one to three mutations in yeast alcohol dehydrogenase are known (from selection experiments) to alter the kinetics (binding constants and cooperatives) only slightly, but sufficiently to alter the equilibrium between normal and slightly structurally different, but harmful substrate (Wills, 1976). Presumably the mutational changes either produce a structural change which is distributed over the whole molecule, including the sites responsible for pattern recognition and cooperativity, or they produce more significant structural change in regions whose only function is to buffer the structural change in the critical region.

The above remarks strongly suggest that gradualism, so important for evolution, is itself a trait susceptible to selective evolution. Indeed, if this were not the case, it would not be justified to retain the assumption that tertiary structure and function are uniquely determined by primary structure (since at least one aspect of function would not be so determined). What remains to be shown, however, is that gradualism is itself gradually evolvable; for otherwise the problem of the improbability of numerous simultaneous mutations would again arise and the adaptive landscape could not be expected to be optimizable in a reasonable time frame.

There are undoubtedly numerous ways of adding redundant weak bonding of the type which makes gradualism possible. This can always be done by increasing the free energy of folding through the addition of amino acids which are functionally superfluous except for their contribution to redundancy or through replacements which strengthen given weak interactions. Such additions or replacements are single step processes, which means that gradualism can always be increased in the gradualistic mode. In principle, rapid evolution of gradualism would also be possible to the extent that this can be mediated by nonspecific addition of amino acids of certain types, as opposed to requiring specific amino acid formats. As discussed in the previous section, the lengthening of the primary sequence resulting from additions of either type does not materially interfere with the rate of evolution and therefore would not be significantly selected against on this ground. Also note that the evolvability of gradualism argument can be turned around and treated as a logical requirement for acceptable rate of evolution (since for this gradualism is a fundamental requirement and degree of gradualism is a potentially variable feature of protein organization).

Why assume that the condition for evolution is that slight changes in primary structure are often concomitant to slight changes in protein function when all that is necessary is that they are concomitant to increasing fitness, perhaps involving very different functional properties? There are an enormous number of possible primary sequences (e.g. 20^{300} for a three hundred amino acid protein) in comparison to the actual number which could reasonably be expected to be associated with a useful or potentially useful function or which could even be expected to fold uniquely. This means that the number of functionally viable proteins is very sparse in the total number of proteins, implying that the requirement of continuity of fitness with at least some step by step changes in sequences would be unfulfillable except under highly special conditions. The plausible special condition is continuity of function change concomitant to continuity of shape change.

The conclusion from this section is: any mutation which improves the degree of gradualism and therefore the conditions for evolution will allow more rapid optimization of fitness (with regard to traits which actually contribute to the organism's function as opposed to the evolution of this function) and therefore will accrue a selective advantage (potentially counteracted only by the selective disadvantage of the extra energy requirements concomitant to larger proteins). In terms of the adaptive landscape there are always smooth paths which lead to neighborhoods (in general of higher nondegenerate dimensionality) where pathways are even smoother and perhaps more numerous. Populations are drawn along these smooth paths, for with each move closer to the optimal landscape they climb more rapidly those particular paths which lead to a maximum of fitness. Darwinian evolution inevitably drives biological systems to regions of the adaptive landscape where Darwinian evolution works best.

5. Implications

The argument in the previous sections can be summed up in a very general bootstrap principle, to be called the principle of molecular adaptability. This is: <u>the gradualness with which protein function changes with single changes in primary structure determines the amenability of the adaptive landscape to evolutionary hill climbing and is itself an evolved property which is gradually adjustable in the course of evolution</u>. This principle has a number of implications at the level of the gene and enzyme, and by extension of the argument, at higher levels of genetic and phenotypic organization. Of particular interest are:

(i) <u>Interpretation of molecular morphology</u>. The relation between structure and function in proteins must be interpreted in terms of their capacity to undergo evolution as well as in terms of their biochemical and physiological function (e.g. rate constants under various conditions). Much of this structure must be entirely superfluous from the strictly biochemical point of view. We note that necessity for considering amenability to evolution is not the same as saying that the actual structure of the molecule is partly a historical property and therefore understandable only in phylogenetic terms (which is undoubtedly true). Amenability to evolution means in particular that certain amino acids and molecular formats subserve primarily the function of allowing gradated distortion of the molecule with discrete change in sequence.

(ii) Isoenzymes and the issue of neutralism. The discovery of unexpectedly large numbers of enzymatic variants of enzymes, either coded at the same or different loci, has raised questions about the role of selection in evolution and in particular the possibility of selectively neutral evolution. Since evolved proteins necessarily occupy regions of the adaptive landscape in which slight changes in primary structure are likely to give rise to slight changes in function, it is inevitable that numerous variations of any given protein will either make the same or a similar contribution to fitness. This does not mean that selection is unimportant, but to the contrary that some degree of neutralism or near neutralism is the inevitable concomitant of the conditions under which selection can effectively act. Thus the principle of molecular adaptability is relevant to interpreting not only the morphology of proteins, but also the statistical distribution of variant types (i.e. isoenzymes). More important, it provides an explanation for the phenomena of neutral selection theory (sometimes misnamed non-Darwinian evolution) in terms which are consistent with and, in conjunction with molecular considerations, implied by classical, selectionist theory.

The above argument should not be taken to imply that the occurrence of isoenzymes cannot frequently be associated with subtle but functionally significant modulations of molecular function. The type of gradualism which allows change of function through evolution also makes it possible for the statistical distribution of enzymes which perform given biochemical or physiological functions to perform these functions optimally under slightly different conditions, thereby subserving a homeostatic function. Some recent models of immunity (cf. Jerne, 1955) and learning (cf. Conrad, 1973, 1974b, 1976) also postulate a fundamental physiological role for ontogenetic processes analogous to natural evolution and therefore for the gradualism property. It should also be noted that increase in protein size concomitant to increase in the free energy of folding may subserve the maintenance of the three dimensional structure of proteins in extreme environments, e.g. in bacteria capable of living under unusually high temperatures.

(iii) Strategies of molecular adaptability. In principle optimal adjustment of the gradualism property would mean that single amino acid changes would be accompanied by a maximal increase in fitness--clearly increasing the gradualism too much would only have the effect of inserting extra steps between an original protein shape and an alternate, similar shape of higher selective value. However, it is clearly impossible for evolution to precisely optimize gradualism, for no general statement can be made about the selection forces in different cases, about the degree of desired shape change given these selection forces, or the degree of shape change that would result from any single amino acid change (which would certainly vary with the evolution of the molecular morphology). In this sense one cannot properly speak of strategies of molecular adaptability, but rather must regard the evolved degree of gradualism as a condition for the past evolution of the molecule and not necessarily optimally suited for its future evolution (except insofar as conditions for past success serve as a valid indication of conditions for future success). It might reasonably be concluded that evolved degree of gradualism varies considerably among the various different enzyme types and that the selective influences acting on it depend on the relative importance of other modes of adaptability (e.g. developmental and behavioral plasticity, cf. Conrad, 1975, 1977).

In principle it is possible for selection to reduce as well as increase degree of gradualism, for if increase is amenable to evolution, so is decrease. This has interesting implications. The gradualism property (at the molecular level) might reasonably be presumed to have been most important early in the development of life, during the age of basic biochemical evolution and adaptive radiation of enzyme types accompanying this evolution. High molecular adaptability allows for protein evolution through the transfer of function mechanism, i.e. through gradual changes in enzyme shape which enable an enzyme species catalyzing reaction A to evolve through a sequence of species each catalyzing in different degree two reactions, A and B, and finally into a species catalyzing only B and not A at all. A high degree of molecular adaptability is costly in terms of energy (e.g. in terms of extra amino acids), implying that once the possibilities for radiation are exhausted, those species (or organisms) which forgo some degree of gradualism may have an advantage. Moreover, such reduction in molecular adaptability has the potential advantage that it provides a mechanism for locking in already developed molecular adaptations. The prediction is that this type of evolutionary conservativism would be most important for the most critical enzymes, viz. those involved in the coding process and that these should therefore be relatively less adaptable in the evolutionary sense. An amusing possibility is that in ancient times, during the age of basic biochemical evolution, many enzyme species were very much larger than species which subserve the same biochemical function today.

6. Extensions to Multigene and Ecological Systems

Attention has been focused on the evolution of a single protein sequence (or a single gene). This is fundamental since protein catalysis ultimately underlies the behavior of biological systems and therefore it could hardly be expected that evolution would be possible at all in the absence of amenability of individual genes to evolutionary change. Also, attention has been focused on point mutation with not too much concern for recombination and crossing over. However, these are also probabilistic processes, unlikely to occur simultaneously. Their importance is not that they make it possible to bypass step by step evolution, but that they bring together different independently evolved pieces of the gene in single steps--in short they make it possible to utilize hierarchy to take better advantage of continuity. Within the single gene, however, the importance of such piecewise evolution would seem to be limited by the extent to which the folded shape of the pieces influence one another. Another mechanism of possible importance are "scratch space genes," i.e.

redundant genes used for low cost evolutionary search (cf. Glaser, 1971). In this case the evolution process is freed from the requirement that it proceed through changes concomitant to similar or increasing fitness. Such drift-like search may play a role in some cases, but its importance is reduced by the fact that correlation between sequence and population size is fortuitous until a significant effect on fitness is established.

The single gene clearly evolves in the context of numerous other genes with which it must co-function and to which it must be coadapted. This is why the argument has been formulated in terms of the contribution of change in primary sequence to change in fitness of the organism rather than in terms of the fitness of single protein species. Thus the evolution of any gene must be compatible with the remainder of the gene complex and also with the evolution of this complex. Furthermore the argument about the relative speed advantage of stepwise as compared with simultaneous modes of evolution is equally valid for sequences of genes as for sequences of amino acids in a single gene (again within the framework of hierarchical processes such as recombination and crossing over which make it possible to put together independently involved subsequences of genes).

The question for individual genes was, what organization allows the proteins (and possibly some other macromolecules) for which they code to have the gradualism property? The same question can be asked about sequences of genes. Clearly gradualism in this case cannot be based directly on folding--for it is the individual proteins which fold. One point is that the irrelevance of folding to the genome as a whole means that order in the sequence is not in general so important, which expands the possibilities for recombination and crossing over. More important, however, is quantitative inheritance. In this case gradual modifiability is based on the additive effect of attenuated enzymes coded by redundant genes. These redundant genes thus play a role analogous to the redundant weak bonding within single enzymes, except that now gradualism is based on change in enzyme concentration rather than change in enzyme shape (but both affecting the rate of processes). As with shape-based gradualism at the molecular level, concentration-based gradualism involves cost in terms of energy (more genes and in general greater numbers of enzymes) and in terms of complexity of reaction structure and control. Once appropriate concentrations evolve they can be fixed by using inversion or other structural properties of the chromosome to effectively block gradual change (analogous to gradualism reduction at the individual molecule level). The possibility for quantitative, concentration-based gradualism ultimately rests on molecular adaptability since enzymes with appropriate rate constants must evolve in the first place. A plausible hypothesis is that adaptive radiation within given taxonomic categories principally involves gradual variations on basically identical developmental plans, where the variation either is mediated by quantitative changes in gene number or gradual quantitative changes in single genes. The resulting ontogenetic processes are simulations of one another in the sense that they can be regarded as deformations in the realization of a prototype developmental scheme (Conrad, 1974b). This is consistent with the topological transformations of Thompson (1917) and seems to be required from the standpoint of the rate of evolution.

Another simplification is that the structure of the landscape has been considered constant--only the topology changes and this is due to movement along the smooth direction (associated with redundancy of weak bonds in the case of concentrations). Globally this structure has many peaks, corresponding to different possible niches. Thus from the standpoint of the ecosystem it is appropriate to picture active hill climbing on many of these peaks simultaneously. A condition for the stability of an ecosystem is clearly that the hill structure remains constant, i.e. that the selection forces and therefore fitness values do not change with time and in particular that the peaks do not move as a result of being climbed. The condition for the nonoccurrence of an extremely damaging instability is that the hills do not in general move faster than the climbers. This is important, for it means that selection for rate of evolution is not based solely on the advantages accruing to those species which can optimize more rapidly but also on the sharp, virtually threshold disadvantage of "de-optimization" in those species which cannot keep up with the landscape. Thus selection for both molecular and quantitative adaptability is necessarily intense except under conditions which are relatively stable and in which species appropriately fitted to new, rapidly emerging hill structures are not present in latent form or cannot immigrate from some other ecosystem (as must have been the case at some stage of evolution).

7. Conclusions

The basic argument advanced in this paper is that smoothness of pathways in the adaptive landscape is important for evolution, that the degree of smoothness is determined not only by the selection forces (i.e. environment) but also by biological organization, and that as corollary to this degree of smoothness is itself an evolvable property. The central question is, what forms of organization smooth the landscape? At the single gene level the answer is that increased free energy of folding concomitant to increased redundancy of weak bonding and also utilization of amino acids with a greater number of analogs allows for gradual distortion of shape with stepwise change in primary sequence and therefore gradual change in specificity. At the polygenic level the answer is that increased redundancy coding for either smaller numbers or less potent enzymes allows for gradual change in enzyme concentration and therefore in the rates of various processes. Both types of smoothness are themselves amenable to evolutionary optimization, but only in the limited sense that degrees of smoothness selectively favored in the past are likely to be appropriate in the future. Of the two types of smoothness, mechanisms at the molecular level are more fundamental since quantitative gradualism is finally contingent on the evolution of appropriate molecular qualities. The rate of evolution advantage of smoothness increases more sharply as systems approach more closely an organization which allows for the single step mode of evolution and also for maximum increase of fitness with each step. Selection for smoothness has threshold intensity if the structure of the landscape is changing in such a way that further evolution is required, but the landscape has not been sufficiently smoothed (by previous evolution) for species to keep up with it. While smoothing the landscape does not ensure that every or even a small minority of peaks in the global landscape will be found, it does ensure that once they are found they will be ascended to the top and relatively rapidly. If the structure of the landscape is changing, new possibilities are opened for pulling species into the domains of attraction of previously unfound peaks or for peak splitting and speciation.

A fundamental point is that smoothing the landscape requires free energy (e.g. synthesis of large proteins in the case of molecular adaptability). However, once this free energy investment is made, the number of trials typically required to reach the top of an adaptive peak decreases and relatively more time is spent higher on the peak. This means that the average fitness of the population is greater than it would otherwise be, in part because the search process is less dissipative (since fewer trials mean fewer misfit forms) and in part because spending more time at higher levels (or less time at lower levels) earns a free energy advantage. Thus the free energy investment in terms of organism function required to smooth the adaptive landscape is compensated by a decrease in the free energy loss associated with the evolution process (as well as an earlier gain accruing from any free energy advantage resulting from this process). More generally, the increase in dissipation concomitant to smoothing is complementary to a decrease in dissipation concomitant to evolution.

References

Aho, A.V., J.E. Hopcroft, and J.D. Ullman: The Design and Analysis of Computer Algorithms, Addison-Wesley, Reading, Mass., 1974.

Barricelli, N.A.: Numerical testing of evolution theories, I and II, Acta Biotheoretica 16, 69 (1962).

Bremermann, H.J.: Optimization through evolution and recombination. In: Self-Organizing Systems, ed. by Yovits, Jacobi and Goldstein. Spartan Books, Washington D.C., 1962.

Bremermann, H.J.: Quantitative aspects of goal-seeking self-organizing systems. In: Progress in Theoretical Biology, vol. I, ed. by F. Snell, pp. 59-77. Academic Press, New York, 1967a.

Bremermann, H.J.: Quantum noise and information, Proc. Fifth Berkeley Sympos. Math. Stat. and Prob. IV, pp. 15-20. University of California Press, Berkeley, 1967b.

Conrad, M.: Computer Experiments on the Evolution of Coadaptation in a Primitive Ecosystem, Ph.D. Dissertation, Biophysics Program, Stanford University, 1969.

Conrad, M.: Statistical and hierarchical aspects of biological organization. In: Towards a Theoretical Biology, ed. by C.H. Waddington, pp. 189-221. Edinburgh University Press, Edinburgh, 1972.

Conrad, M.: Is the brain an effective computer? Intern. J. Neuroscience 5, 167-170 (1973).

Conrad, M.: Evolutionary learning circuits, J. Theoret. Biol. 46, 167-188 (1974a).

Conrad, M.: Molecular automata. In: Physics and Mathematics of the Nervous System, ed. by Conrad, Guttinger, and Dal Cin, pp. 419-430. Springer-Verlag, Heidelberg, 1974b.

Conrad, M.: Analyzing ecosystem adaptability, Math. Biosciences 27, 213-230 (1975).

Conrad, M.: Complementary models of learning and memory, Biosystems 8, 119-138 (1976).

Conrad, M.: Functional significance of biological variability, Bull. Math. Biol. 39, 139-156 (1977).

Conrad, M. and H.H. Pattee: Evolution experiments with an artificial ecosystem, J. Theoret. Biol. 28, 393-409 (1970).

Friedberg, R: A learning machine, IBM Journal of Research and Development, vol. 2, 2-13 (1958).

Glaser, D.: Lecture, University of California Berkeley, 1971.

Holland, J.: Adaptation in Natural and Artificial Systems, University of Michigan Press, Ann Arbor, 1975.

Jerne, B.: The natural selection theory of antibody formation, Proc. Nat. Acad. Sci. U.S. 41, 849-857 (1955).

Karp, R.M.: Reducibility among combinatorial problems. In: Complexity of Computer Computations, ed. by Miller and Thatcher, pp. 85-104. Plenum Press, New York, 1972.

Landauer, R.: Fundamental Limitations in the Computational Process, IBM Thomas J. Watson Research Center Preprint, 1976.

Mayr, E.: Animal Species and Evolution, Harvard University Press, Cambridge, Mass., 1963.

Perutz, M.F.: Proteins and Nucleic Acids, Elsevier, Amsterdam, 1962.

Rechenberg, I.: Evolutionsstrategie, Optimierung technischer Systeme nach Prinzipien der biologischen Evolution, Friedrich Frommann-Verlag, Stuttgart-Bad Canstatt, 1973.

Reed, Toombs, and Barricelli: Simulation of biological evolution and machine learning, J. Theoret. Biol. 17, 319 (1967).

Roitt, I.: Essential Immunology, 2nd ed., Blackwell Scientific Publications, Oxford, 1974.

Rossman, M.G. A. Liljas, C. Branden, and L.J. Banaszak: Evolutionary and structural relationships among dehydrogenases. In: The Enzymes, 3rd ed., vol. 11, ed. by P.D. Boyer, p. 61. Academic Press, New York, 1975.

Simpson, G.G.: The Meaning of Evolution, Yale University Press, New Haven, 1949.

Thom, R.: Topological models in biology. In: Towards a Theoretical Biology, vol. 1, ed. by C.H. Waddington, pp. 95-120. Edinburgh University Press, 1970.

Thompson, D'Arcy Wentworth: On Growth and Form, Cambridge University Press, 1917.

Waddington, C.H.: The basic ideas of biology. In: Towards a Theoretical Biology, vol. 1, ed. by C.H. Waddington, p. 19. Edinburgh University Press, 1968.

Wills, C.: Production of yeast alcohol dehydrogenase isoenzymes by selection, Nature 261, 26-29 (1976).

Wright, S.: The roles of mutation, inbreeding, cross-breeding and selection in evolution, Proc. Sixth Intern. Congr. Genet. 1, 356-366 (1932).

APPENDIX

To show that F(m) decreases with m for all realistic values of p rewrite Eq. (7) as

$$F(m) = A + B \tag{A1}$$

where

$$A = 19^{1-m} m p^{m-1} (1-p)^{1-m} \tag{A2}$$

and

$$B = \frac{(m-1) D\ N_o p^m (1-p)^{n-m}}{19^m} \tag{A3}$$

It is only necessary (treating m as continuous) to show that $\partial A/\partial m < 0$ and $\partial B/\partial m < 0$ for all realistic (and even unrealistic) values of p. Considering A first,

$$\partial A/\partial m = \left[19^{1-m} m p^{m-1} (1-p)^{1-m}\right]\left[\ln p - \ln(1-p) - \ln 19 + 1/m\right] < 0 \tag{A4}$$

provided that the expression in the second bracket is less than zero (since the expression in the first bracket is always positive). This is the case as long as

$$\ln \frac{19 - 19p}{p} > 1/m \tag{A5}$$

which is true for all realistic values of p (i.e. $p < .86$ in the worst case). Calculation of $\partial B/\partial m$ would show that the B term exhibits the same general behavior, implying that F(m) increases with m for all realistic values of p. However, it is more instructive to observe that for all values of p and n such that

$$\frac{1}{p(1-p)^{n-1}} > \frac{(m-1) D\ N_o}{19m}, \tag{A6}$$

A dominates B and in fact B in general makes a negligible contribution to F(m). This is certainly the case for $p < 10^{-6}$ and $n > 100$, which is why the values of F(m) are so insensitive to D and N_o in the tables.

To show that the "second derivative" behavior of Table I is general it is sufficient to show

$$\partial^2 A/\partial m^2 = \left[19^{1-m} m p^{m-1} (1-p)^{1-m}\right] \left[-m^{-2} + (\ln p - \ln(1-p) - \ln 19 + 1/m)(\ln p - \ln(1-p) - \ln 19 + 1)\right] > 0 \qquad \text{(A7)}$$

since the only values of p of interest are those for which condition (A6) holds. The expression in the first bracket of (A7) is always positive and therefore it is only necessary to show that the expression in the second bracket is in general positive, i.e. that

$$(\ln p - \ln(1-p) - \ln 19 + 1/m)(\ln p - \ln(1-p) - \ln 19 + 1) > 1/m^2 \qquad \text{(A8)}$$

However, this holds as long as condition (A5) is met (with m=1 for the expression in the second brackets) and therefore for all reasonable values of p the slope of F(m) is negative but decreasingly negative as m increases. Note that 1/F(m) is the ratio of simultaneous to stepwise evolution times (or, alternatively, the ratio of the frequencies of stepwise to simultaneous type changes) and therefore can be identified with the selective advantage of the stepwise mode.

CELL ASSEMBLIES IN THE CEREBRAL CORTEX

V. Braitenberg
Max-Planck-Institut für biologische Kybernetik
Spemannstrasse 38
74 Tübingen

I. INTRODUCTION

To say that an animal responds to sensory stimuli may not be the most natural and efficient way to describe behaviour. Rather, it appears that animals most of the time react to situations, to opponents or things which they actively isolate from their environment. Situations, things, partners or opponents are, in a way, the terms of behaviour. It is legitimate, therefore, to ask what phenomena correspond to them in the internal activity of the brain, or, in other words: how are the meaningful chunks of experience "represented" in the brain?

A crude version of this question takes the form: is the presence of a relevant happening signalled by the activity of just one neuron, which otherwise is always silent, or is it represented by an irreducibly complex description of the activity of the brain?

Following an old idea (Hebb, 1949 and older) we shall explore the possibility that it is neither single neurons nor abstract diffuse properties of the state of the brain which correspond to the relevant events of behaviour, but something in between, identifiable sets of neurons.

These "cell assemblies" have recently gained support from neurophysiology in two ways. First, many years of recording responses of single neurons to sensory stimuli have shown that no very complicated or very unique input is needed to activate a neuron. The most efficient stimuli for cortical neurons are rather elementary configurations of the sensory input, such as moving lines in narrow regions of the visual field (Hubel and Wiesel, 1959) or changing frequencies in certain delimited regions of the acoustic spectrum (Evans, 1968). These "features" cannot independently carry meaning but must be in the same relation to meaningful events as the phonemes of linguistics are to words or sentences. The whole meaningful event must be signalled in the brain by a set of neurons, each contributing a particular aspect which that event may have in common with many other events.

The second line of evidence is derived from the neurophysiology of learning. It was one of Hebb's points that cell assemblies representing things in the brain are held together by excitatory connections between the neurons of which they are composed, and that these connections are established through a learning process. The most natural way in which such learning could take place is if a statistical correlation, say, a frequent coincidence of a certain set of elementary features in the input were transformed into synaptic connections between the corresponding neurons. Some recent observations on the plasticity of the connections of single neurons (Hubel and Wiesel, 1965; Wiesel and Hubel, 1965; Blakemore and Cooper, 1971; Hirsch and Spinelli, 1971) can indeed be explained by invoking such a mechanism.

It seems timely, therefore, to reconsider cell assemblies as a possible substrate of behaviour, and, particularly, to review the cerebral cortex with the idea in mind that it might be the place where cell assemblies are formed and sustained.

There remains one nagging thought, however, when we dismiss single neurons as the elements onto which complex situations, things, etc. are mapped in the brain. Suppose we were recording with a microelectrode from a neuron whose activity corresponds exactly to the presence of a close relative, or to a particular bird's song, or to the memory of a particular scent. The probability that, while we are recording we stumble accidentally upon such very rare stimulus configurations is negligeably small. Thus, strictly speaking, the idea of single neurons as classificatory elements for complex situations is beyond proof or disproof. Of course for similar reasons and for added technical difficulties it is even more hopeless to expect experimental proof of the correspondence of a cell assembly to a certain event in the input. In this field we must rely on theoretical plausibility and on considerations about the structure of networks in the brain that may seem more suited for one or the other scheme.

II. CELL ASSEMBLIES

Terminology. A cell assembly is a set of neurons interconnected by excitatory synapses. If we want to be more explicit, several definitions are possible. We found the following rather convenient: a cell assembly is a set of neurons, each of which receives excitation from and gives excitation to some other members of the same set. Moreover, a cell assembly is a collection of neurons which cannot be cut into two separate collections without severing at least two excitatory fibers, one for each direction.

By this definition the neurons of the halo of the cell assembly are not included, namely those neurons that receive excitation from the cell assembly (and thus are always active together with it) without, however, contributing excitation to it.

Neither are the afferents of the cell assembly included, which give it excitation without receiving excitation in turn.

It is not necessary that the cell assembly be completely connected, i.e. that each component cell send a fiber to each of the others, a condition which, as we shall see, is hardly compatible with realistic assumptions.

A cell assembly may contain sub-assemblies for which the same definition holds. Actually the entire nervous system, minus motor and sensory neurons (provided that they are not included in feedback loops) is a cell assembly of which all the others are sub-assemblies.

Neurons have thresholds of excitation for which they become active. We suppose that these are under the influence of an external threshold control. In terms of real neurons, it is irrelevant whether this is due to a background excitation or inhibition of the neurons, or to a true control of the thresholds (the latter being, however, the least realistic assumption).

We say "a cell assembly holds at a threshold $\leq \Theta$" (or simply: at threshold Θ) when at that threshold all the neurons of the assembly, once excited, stay active due to their reciprocal excitatory connections.

A cell assembly holding at threshold Θ may have sub-assemblies that hold at the same or at lower or at higher thresholds.

Improper cell assembly we may call a collection of neurons whose excitation will hold at a certain threshold because each is part of a proper cell assembly but not all of the same. Improper cell assemblies lack a fundamental property of cell assemblies, namely that for thresholds low enough, partial excitation of the cell assembly may spread to (ignite [+)]) the entire assembly.

A cell assembly is homogeneous when each component neuron gives (and receives) excitation of a certain strength to (from) the same number of other neurons of the same assembly. If it is non-homogeneous, it may contain sub-assemblies which are more strongly connected than the rest (centers). We may distinguish monocentric and polycentric non-homogeneous cell assemblies, depending on whether at higher thresholds one or several disjunct sub-assemblies will hold. Cell assemblies are disjunct when they do not share any neurons.

In the following the reader will notice that the discussion which has up to now been investigating the logical consequences of known facts about the brain, acquires more the character of an uninhibited invention of mechanisms that might be useful in a functioning brain. This is in accordance with an attitude expressed by Palm and Heim in the introduction to this meeting (and by Rosenblueth, Wiener and Bigelow, 1943) which would accept the concept of design as a legitimate tool in the analysis of biological complex systems. The model which emerges from these considerations will then be tested for its plausibility in the last part of this paper, which is about numbers of neurons and synapses in the cerebral cortex.

Dynamics of threshold control. It is necessary that the thresholds of the neurons be under the influence of a feedback from

---

[+)] I owe this term to Charles Legendy

the activity of the whole network for the following reasons: a) in order to prevent the extension of the activity as an epileptic fit to the maximal cell assembly, that of the whole brain; b) to keep the activity above a minimal level; c) in order to distinguish proper from improper cell assemblies; d) in order to discover centers within a cell assembly. We have to think of cell assemblies in a wider context encompassing an external control apparatus in addition to the matrix of neurons in which they occur (Fig. 1).

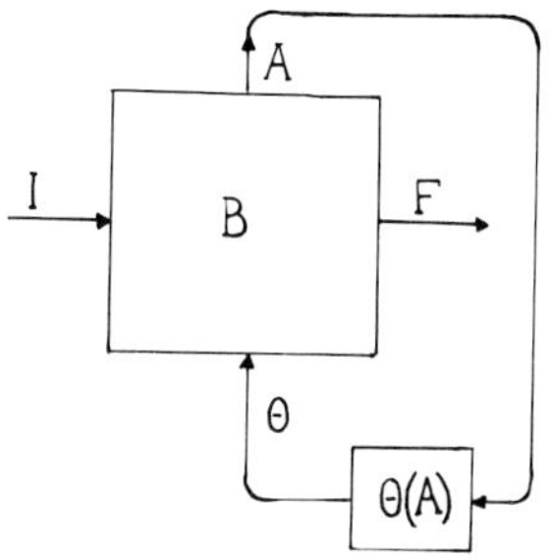

Fig. 1

Given an input I, which excites some of the neurons, we obtain an output F of active neurons which depends also on the cell assemblies that have ignited. Their ignition in turn depends on the threshold Θ, which is set by the threshold control mechanism $\Theta(A)$ depending on the total activity A of the network. The simplest definition of A is the number of active neurons within the box B. There is an ample choice for the form of the function $\Theta(A)$. This may be simply a suitable monotonically increasing function of A, or Θ may depend on the variation of A with respect to time. The most interesting possibilities are offered by a dependence of the form $\Theta(\frac{\Delta A}{\Delta\Theta})$ (Fig. 2). We observe the activity A, i.e. the number of active neurons, as a function of the threshold Θ, which we suppose to be the same for all neurons. The neurons are under the influence of a certain constant input, which provides various strengths of excitation for each of them. Suppose we lower the threshold slowly starting from a value which is higher than the excitation of the most excited neurons. The activity will increase and it might be desirable to set the threshold at a value corresponding to one of the plateaus Θ_1 or Θ_2 of the

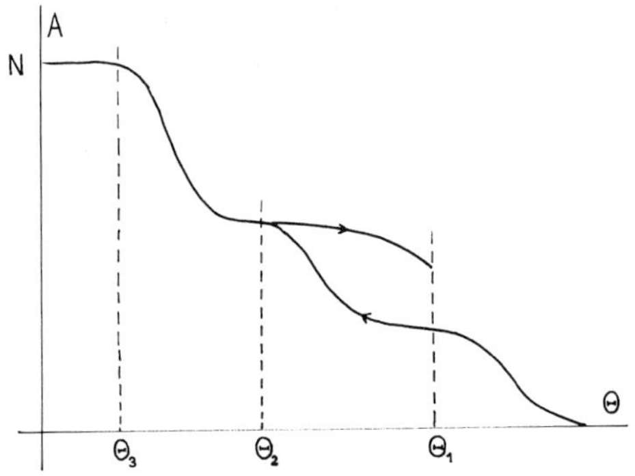

Fig. 2

curve $A(\Theta)$. This will tend to produce activity in sets of neurons which have a similar relation to the input. We will avoid Θ_3 beyond which the activity goes over into the maximal, seizurelike state with all N neurons active. If we lower the threshold and then raise it again, say between Θ_1 and Θ_2, we may find hysteresis of $A(\Theta)$ if at Θ_2 some cell assembly has ignited which raises the level of internal excitation so that the activity will stay at a higher level when the threshold is moved back to Θ_1. Thus it might be desirable in addition to provide the threshold control mechanism with a certain strategy $\Theta(t)$ which may be used to discover such intervals of the dependence $A(\Theta)$ where the divergence of the two branches of the hysteresis curve for small variations of the threshold is largest. By this means the ignition of cell assemblies related to a certain input could be discovered.

The idea of a threshold control based on the activity of the whole network makes it possible to think of ensembles of neurons with exclusively excitatory connections. Or rather, the inhibitory interactions which are necessary for stability, are all relegated to an external mechanism. This is not realistic even for the cerebral cortex, but, as we shall see, it helps to portray this structure from an angle that may reveal one of its salient characteristics.

Sequences of cell assemblies. It is not difficult to invent a threshold control $\Theta(A)$ and a strategy $\Theta(t)$ such that, for a constant input, the ignition of several different cell assemblies will occur one after the other. They may be characterized by different thresholds, or by different approaches to one and the same threshold which have different effects because of the hysteresis characteristic of $\Theta(A)$. One may express this by saying that the system hunts for an interpretation of the input, or that it "thinks". The input provides a certain distribution I of excitation on the neurons of the system. A convenient threshold will transform this into a configuration of active neurons FI. These will in turn contribute excitation whose distribution we may call EFI. If the input remains constant, the threshold will now produce a set of neurons F (I + EFI) which will in general be different from FI and may not even include the neurons of FI. There will be a new distribution of excitation, etc. A final state may eventually be reached which we may call F^*I.

A periodic operation (colloquially called the pump of thoughts) may involve the following steps. Given a certain input I, the threshold is lowered so that the set of active neurons FI will go over into a larger set F'I. This will encourage the ignition of cell assemblies. As the threshold is again raised, activity is smothered and only the most strongly connected cell assembly will survive. A new cycle beginning again with a lowered threshold will bring in new cell assemblies. They may include an even more strongly connected cell assembly, which will be the next one to survive when the threshold is raised. The evolution will be in the direction of the most strongly connected cell assemblies.

Such sequences of cell assemblies are more interesting when one assumes adaptation or fatigue of the neurons, such that neurons participating in an active cell assembly will subsequently have higher thresholds for some time. This assumption is quite realistic, since slow variations of the excitability of neurons as a consequence of activity have been quite commonly observed. Adaptation would tend to make persisting cell assemblies fade and would favour fresh cell assemblies in the pumping process described.

Temporal structure of cell assemblies. The segmentation produced by a periodicity in the threshold control defines unitary episodes of neuronal activity which we may call syllables, borrowing a term from linguistics. Clearly, each of these may again show variation within its duration. The ignition of a cell assembly takes time, the threshold does not stay constant, and, more important still, there may be various modes of oscillation within a cell assembly, possibly different modes within the same cell assembly depending on the way excitation reached it. It is quite easy and quite realistic to imagine alternating sets of active neurons within a cell assembly, cyclic activity etc.

Interpretation. If single neurons in a sensory area of the brain correspond to individual features or properties of things in the sensory space, it is reasonable to suppose that connected sets of neurons, i.e. cell assemblies correspond to the things themselves. The excitatory synapses within the cell assembly are an image of the logical implications that tie the properties into a coherent set of properties, or of the forces that bind the various physical components of that thing. Outside of the direct sensory representations we may expect cell assemblies to represent bundles of more abstract properties. The fact that each cell assembly holds at a certain threshold corresponds to the binary character of the recognition of a thing. The observation that with lower thresholds, the excitation of part of the set of neurons may suffice to ignite the whole assembly reflects the well known phenomenon of the completion of figures in perception: some of the properties of a thing make us perceive the whole thing under certain circumstances. In this respect cell assemblies behave exactly like the "threshold devices" or "formal neurons" of the old "calculus of ideas" (McCulloch and Pitts, 1943).

The chaining of cell assemblies that we described as a consequence of periodic variations of the threshold is what is called a chain of associations in psychology. The halo of an active cell assembly represents the "consequences" of the thing represented, and will determine the next cell assembly that will ignite in the sequence. There is an important difference in the interpretation of excitatory synapses within the cell assembly on one hand and those that carry excitation from the cell assembly to the halo on the other. The former represent the symmetrical relation of belonging together, while the second embody the asymmetrical relation of consequence, temporal sequence or causality.

III. THE CEREBRAL CORTEX

I have recently developed some thoughts on the role of the cerebral cortex, based mainly on considerations about the number of input, output and internal elements and of their synaptic connections (Braitenberg, 1974, 1977). I shall review them here from the point of view of cell assemblies.

The cortex appears as a massive system of connected neurons, greatly outnumbering (by a factor of the order 10^3) the neurons carrying information to the cortex from the sensory system and those connecting the cortex to the motor output.

This characterizes the cortex among other, equally impressive cerebral analyzers of sensory information: the optic tectum of the frog, e.g., contains about as many neurons as there are fibers feeding it visual information. There are good reasons to consider the most numerous neuronal cell type, the pyramidal cells, as the basic neuronal equipment of the cortex. It is not entirely certain, but it is assumed here that the connections between the pyramidal cells are excitatory. The reasons for this assumption are the following:

a)The cerebral cortex (and especially the hippocampal region) is the piece of nervous tissue most susceptible to epileptic activity (Jasper, 1969). If enough neurons are activated, the most diverse stimuli can produce self sustained seizure-like activity in the cortex. One way of doing this is to make an electric current pass through the tissue. This presumably excites indiscriminately excitatory as well as inhibitory neurons. The fact that a seizure ensues, shows that the excitatory connections prevail over the inhibitory ones. It is reasonable, then, to make the pyramidal cells, the most numerous cell type, responsible for the excitatory connections.

b) The fibers of the corpus callosum, which are axons of pyramidal cells, certainly make excitatory connections since they convey epileptic activity from one side of the brain to the other ("mirror focus", Morrell, 1961). Their excitatory nature has also been directly observed by electrophysiological means (Renaud, quoted by Gloor, 1972).

c) The axons of cortical pyramidal cells which reach distant places, e.g. the spinal cord, make excitatory connections.

We may see in the almost 10^{10} pyramidal cells of the human cerebral cortex connected by 10^{14} excitatory synapses the substrate in which cell assemblies develop.

Two sets of connections. The connections between pyramidal cells are collected in two distinct systems of fibers. The fibers of the A-system are the axons of pyramidal cells which traverse the white substance and enter the cortex again in different places in order to terminate (mainly) in the upper layers of the cortex. There they make synaptic connections with the apical dendrites of other pyramidal cells. The B-system consists of the axon collaterals of the pyramidal cell axon which stay within the cortex and make synaptic contact with the basal dendrites of neighbouring pyramidal cells.

The assumption that both the A- and the B-terminals of the excitatory pyramidal cell axon have among their target neurons again pyramidal cells has not been proved directly by electronmicroscopical observation, but is inescapable on quantitative grounds (Braitenberg, 1977; Schüz, 1977). The bulk of the postsynaptic sites is furnished by dendritic spines of pyramidal cells. The larger part of the axonal presynaptic specializations again belong to pyramidal cells. The majority of the afferents of a pyramidal cell must come from pyramidal cells, and vice versa.

Quantitative aspects of the A- and B-system. Even if we assume that the connections between pyramidal cells are ultimately fixed by a process of learning, we should like to know what limits are set to this process by the predetermined anatomy of the cortex. Clearly, there is no room for the 10^{20} connections which would be necessary to provide direct contacts between any pair of pyramidal cells and hence to discover directly every possible coincidence of activity. We shall examine separately the limitations imposed by the anatomy of the white substance (A-system) and by that of the cortical grey substance (B-system).

The white substance of the human brain has a volume of about 10^5 mm^3 on each side. The associational and callosal fibers, cortico-cortical axons of pyramidal cells, make the strongest contribution to this volume. The number of neurons in the thalamus, which provide the strongest system of afferent fibers to the cortex, is of the order 10^8: the afferents cannot be responsible for more than a few percent of the subcortical volume, the efferent motor fibers presumably for even less. Assuming a fiber thickness of 1μm, the average length of the cortico-cortical fibers turns out to be a few centimeters. This length would be compatible with the assumption of random connections from every place in the cortex to every other place. In fact in a flat circular plate, the average length $\bar{l}$ of the connections in such a scheme is slightly less than the radius, $\bar{l} = \left(\frac{16}{3\pi} - \frac{8}{9} \right) r$ (Palm, 1976), in a concave cortex correspondingly less.

The system of cortico-cortical fibers is certainly not randomly organized. Various fiber bundles have been described that connect preferentially certain areas of the cerebral cortex. Also, the corpus callosum is known to connect symmetric places of the two hemispheres (its fibers, which are of the order of 10^8, contribute a few percent of the total cortico-cortical system). However, since the assumption of random connections is compatible with the observed volume of the white substance, it may not be too far from the truth and it is interesting to speculate on the consequences of such a scheme. It seems certain that each pyramidal cell produces only one cortico-cortical fiber. Therefore, if we compartmentalize the cortex into compartments of which it can be said that each is connected to each, each must contain (at least) as many pyramidal cells as there are compartments. A natural parcellation of the cortex therefore would be into $\sqrt{N}$ compartments (N = the total number of pyramidal cells) each containing $\sqrt{N}$ cells. Such compartments in man correspond to columns 1 mm across and contain about 10^5 neurons. In the mouse cortex, analogously, the $\sqrt{N}$ compartments ($N_{mouse} = 10^7$) have a diameter of 0,17 mm. It is interesting that in both species the size of a compartment which could be informed about the state of the whole cortex (assuming, not quite realistically, random connections) is roughly that of the spread of a large pyramidal cell. The idea of $\sqrt{N}$ compartments has an interesting consequence. It is not much information about the state of the cortex if we receive from each compartment only one of the 100.000 fibers emanating from it, unless there is a strong correlation of the activity of the neurons within one compartment. This sheds new light onto the observation of functional columns in the visual, somatosensory and motor regions of the cortex which provide evidence for strong neighbourhood correlation of the activity of cortical neurons.

It is not well known how far the terminal branches of a cortico-cortical fiber spread in the cortex away from the point of re-entry of the fiber. Sporadic observations show rather lose ramifications which seem to give the apical dendrites of many pyramidal cells a chance to contact each fiber. The spread does not seem to be, however, much wider than the width of a compartment.

Thus the connections between pyramidal cells in the A-system are partly predetermined by special fiber bundles which must be responsible for what a species of animals can and what it cannot learn, and to a certain extent by chance. Both principles make the probability of a connection not simply dependent on the distance between two pyramidal cells and therefore justify the term non-metrically dependent connections or simply ametric connections (Palm and Braitenberg, 1977).

On the contrary, the probability of a direct connection in the B-system, i.e. an intracortical connection through axon collaterals, is strongly dependent on the geometric layout of the cortical tissue. These collaterals only span a few hundred microns or at most (depending on the region of the cortex) a few millimeters. Thus, the B-system may be rightly called metric system, or metrically dependent system.

An estimate of the probability of connections between pyramidal cells in the B-system can be obtained as follows (the numbers refer to the cortex of the mouse but may be carefully extrapolated to the human cortex). A comparison of the number of synapses in 1 mm^3 of cortex (10^9) with the sum of the lengths of all axons in 1 mm^3 of cortex (1,2 times 10^9 μm) reveals that each axon makes a synaptic contact almost every micron of its length. Hence, from measurements of the length of all the collaterals of a pyramidal cell the number of synapses for which it is presynaptic can be obtained. This number (a few thousand) matches the number of spines on the basal dendrites of a pyramidal cell, which are probably the sites of the excitatory synapses for which the pyramidal cell is post-synaptic. In the B-system, we may conclude, each pyramidal cell receives and makes a few thousand excitatory contacts. A convenient estimate, corresponding to the number of spines on the basal dendrites of a medium sized pyramidal cell of the mouse cortex, is 2000 synapses. We cannot make much use of this information in connection with cell assemblies, unless two further questions are answered: first, onto how many different neurons the synapses of one pyramidal cell are distributed, (and therefore, from how many different neurons the synapses converge onto one neuron), and, second, how many synapses must be activated in order to meet the threshold of the neuron.

The first question may be answered from anatomy. The axon collaterals of pyramidal cells are peculiarly straight and unbranched (Braitenberg, 1977). The probability of such a fiber hitting the dendritic tree of another pyramidal cell once is quite small, about as small as that of a bullet shot through a leafless tree hitting one of the branches. The probability of hitting it twice is much smaller. A quantitative estimate may be obtained graphically (Fig. 3). We draw the basal part of the dendritic tree of a pyramidal cell, projected onto a plane. For the part of the dendritic tree which is covered with spines we draw an outline encompassing the dendrite plus the spines (thickness 5 μ, = thickness of the dendrites plus twice the length of a spine). The area enclosed by this outline represents the "cross-section" of the dendritic tree as seen by a neuron that sends an axon in a direction perpendicular to the plane of the drawing. We suppose that the axon makes a synaptic contact when it hits the dendritic tree within this area. The probability of such a contact for an axon that enters the region of the dendritic tree is, then, roughly the ratio of the dendritic cross section divided by the area of the circumscribed circle, in the specific case $0{,}0043\ mm^2 / 0{,}038\ mm^2 = 0{,}11$. The shaded area of fig. 3 represents the probability of hitting the dendritic tree twice, $6 \cdot 10^{-4}$. Here the form factor introduced by the clustering of synapses along the dendrites makes the probability for two hits much smaller than the square of that for one hit.

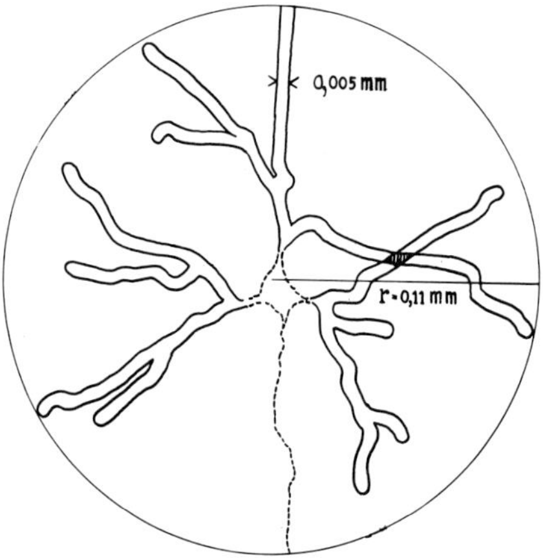

Fig. 3

Another estimate is the following. Within a sphere circumscribed around the basal dendrites of a neuron, that neuron furnishes about 1/1000 of the postsynaptic sites, the others belonging to other pyramidal cells whose dendritic fields are interlocked with the present one. We suppose that the position of these sites in the sphere is random. A fiber finding its way through the sphere for a length of 100 μm makes 100 synapses (n = 100) of which some may belong to the particular neurons in question. If it picks its postsynaptic neurons at random, the binomial distribution will furnish the probabilities w_o, w_1, w_2 ... for no contact, one contact, two contacts, etc. (Fig. 4, top). If all the axon collaterals for a neuron are within the dendritic tree of the other neuron, the number of synapses established may be n = 2000 (Fig. 4, bottom). The table shows that for very close neurons, the most likely connection is through one or two contacts, for more distant ones 1 contact will prevail. More than two contacts with the same neuron are rare according to this reasoning.

Comparing fig. 3 with fig. 4, it is interesting to observe that the two methods give similar values for the probability of one contact in the case of a fiber of a length between 100 and 200 μ

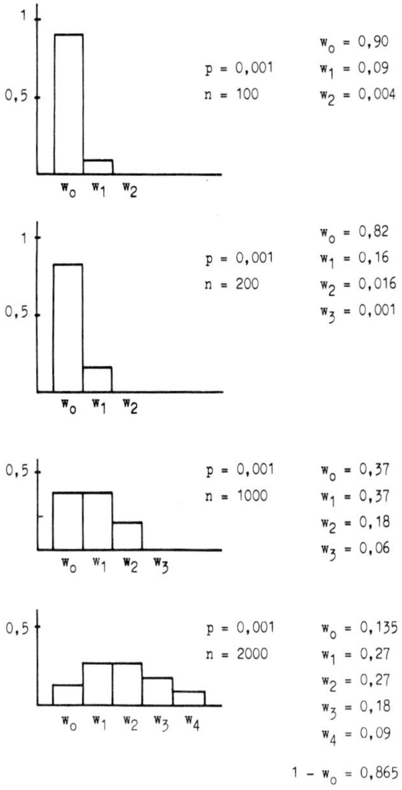

Fig. 4

($w_1 \approx 0{,}1$) and both give very small values for the probability of 3 or more connections.

Thus it seems that the intracortical connections between neighbouring neurons guarantee a great dispersion of the information of one neuron onto many neighbouring neurons, with multiple contacts being the exception rather than the rule.

The A- and the B-system together inform each pyramidal cell of the state of a few thousand other pyramidal cells, some nearby and some scattered throughout the cortex. If the pyramidal cells are the site of the learning process that establishes (or modifies) synaptic connections on the basis of correlated activity, we can see how cell assemblies grow, some locally and some diffuse, involving neurons from disparate regions.

There remains the question of the threshold. A vague estimate is possible on the basis of the phenomenon of the "mirror focus" already mentioned. An epileptic focus on one side of the brain may excite the contralateral cortex so as to produce epileptic activity there, too. Since the callosal fibers are only a few percent of the total cortico-cortical fibers, we shall suppose that for any one neuron they furnish only a few percent of the afferent synapses. Since they excite the neurons to paroxysmal activity, we may conclude that synchronous activation of a few percent of the synapses (say 40 to 80 synapses) is abundantly above threshold.

IV. CONCLUSION

It is worth while to speculate about cell assemblies as an alternative to feature detectors and hierarchies of classificatory units. These concepts are related to Perceptrons. Similarly, cell assemblies would find their technological analogue in a (non existing) Conceptron. Such a device would store correlations between elementary events, and generate the classificatory units when they are needed, in the form of something like "resonant modes" +), i.e. patterns of activity sustained by networks of positive feedback.

The most obvious difference between a brain containing a Perceptron and one making use of a Conceptron lies in the fact that in the first somewhere between input and output the activity may converge onto one or very few lines in order to diverge again in the output, while in the second the activity will always remain distributed in a great number of neurons. The all-or-none characteristic of a "decision" in one scheme resides in the properties of the membrane of the individual neuron, in the other case in the explosive charac-

+) This term, borrowed from physics, may carry misleading associations. Therefore we prefer to talk about cell assemblies as "consonant modes", in spite of possible unwanted linguistic or musical associations, or "conflagrant modes" in accordance with the expression "ignition of a cell assembly".

ter of the positive feedback between neurons.

It would be surprising if it turned out that the real brain makes use only of one or the other scheme. Most likely the two schemes are used in combination, with the hierarchical organization predominating at the sensory and motor periphery of the nervous system, and the cell assemblies in between. From this point of view the cerebral cortex would seem a good place for cell assemblies, and we have seen that it contains the necessary equipment.

The problem of how many "consonant modes" may be established in the cerebral cortex, and how many can be distinguished by means of a shrewd threshold control is a difficult one. The difficulties seem to reside in part in the mathematics, in part in our insufficient psychological knowledge. Just how many concepts does man use? And how many the mouse? How fast do they change with time? (Legendy, 1971).

I believe that the safest data in this field are the anatomical ones, in spite of the great uncertainty which affects some of the measurements.

It is a pleasure to thank Dr. Palm for abundant discussions.

V. REFERENCES

Blakemore, C., and Cooper, G.F.: Modofication of the visual cortex by experience. Brain Res. 31, 366 (1971)

Braitenberg, V.: Thoughts on the cerebral cortex. J. theor. Biol. 46, 421 - 447 (1974)

Braitenberg, V.: Cortical architectonics: General and areal. In: Architectonics of the cerebral cortex. M.A.B. Brazier, H. Petsche, eds., New York: Raven Press 1978, pp. 443 -465

Evans, E.F.: Upper and lower levels of the auditory system: A contrast of structure and function. In: Neural networks. Berlin-Heidelberg-New York: Springer 1968

Hebb, D.O.: The organization of behaviour. New York: John Wiley and son 1949

Hirsch, H.V.B. and Spinelli, D.N.: Modification of the distribution of receptive field orientation in cats by selective visual exposure during development. Exp. Brain Res. 13, 1 - 43 (1971)

Hubel, D.H. and Wiesel, T.N.: Receptive fields of single neurones in the cat's striate cortex. J. Physiol. (Lond.) 148, 574 - 591 (1959)

Hubel, D.H. and Wiesel,T.N.: Binocular interaction in striate cortex of kittens reared with artificial squint. J. Neurophysiol. 28, 1041 - 1059 (1965)

Jasper, H.H.: Mechanisms of propagation: Extracellular studies. In: Brain mechanisms of the epilepsies. H.H. Jasper, A.A. Ward and A. Pope, eds., Boston: Little, Brown and Company 1969

Legendy, C.: A remark on Hebb's cell assemblies. In: Atti del congresso di cibernetica, vol. II. M.A. Baldocchi and F. Lenzi, eds., Pisa: Lito Felici 1971, pp. 905 - 917

McCulloch, W.S. and Pitts, W.H.: A logical calculus of the ideas immanent in nervous activity. Bull. Math. Biophys. 9, 127 - 247 (1943)

Morrell, F.: Lasting changes in synaptic organization produced by continuous neuronal bombardment. In: CIOMS symposium on brain mechanisms and learning. A. Fessard, ed., London: Blackwell 1961, pp. 375 - 392

Palm, G.: private communication, 1976

Palm, G. and Braitenberg, V.: Tentative contributions of neuroanatomy to nerve net theories. In: Third European meeting on cybernetics and systems research 1976, Progress in cybernetics and systems research, vol. 3, London: Advance Publications Limited (in press)

Palm, G. and Heim, R.: this volume

Renaud, quoted by Gloor, P. in: Synchronization of EEG activity in epilepsies. H. Petsche and M.A.B. Brazier, eds., Wien-New York: Springer 1972

Rosenblueth, A., Wiener, N. and Bigelow, J.: Behaviour, purpose and teleology. Philosophy and Science 10, 1943

Schüz, A.: Some facts and hypotheses concerning dendritic spines and learning. In: Architectonics of the cerebral cortex. M.A. B. Brazier, H. Petsche, eds., New York: Raven Press 1978, pp. 129 - 135

Wiesel, T.N. and Hubel, D.H.: Comparison of the effects of unilateral and bilateral eye closure on cortical unit responses in kittens. J. Neurophysiol. 28, 1029 - 1040 (1965)

EXCITATORY AND INHIBITORY PROCESSES GIVING RISE TO THE DELAYED RESPONSE IN THE RETINAL GANGLION CELL OF FROG

D. Varjú
Lehrstuhl für Biokybernetik der Universität Tübingen

1. INTRODUCTION

The response of the ganglion cells in the frog's retina to visual stimuli has been extensively studied during the last four decades. The pioneering work has been done by Hartline (1938, 1940), Barlow (1953, 1957) and by Maturana et al. (1960). Most of the properties of the ganglion cell response can be described satisfactorily on the basis of lateral interactions within the receptive field (RF) of the cell (see rewievs e.g. by Varjú, 1969 and more recently by Grüsser and Grüsser, 1976). One particular response described first by Pickering and Varjú (1967) and termed 'delayed response' is less well understood. This response is elicited mainly by short light flashes projected on to the entire RF or parts of it. The basic properties of the response are illustrated in fig. 1; the results presented there have all been obtained by homogeneously illuminating the entire RF of the cells with flashes of 50 μs halfwidth and of varying intensity I. Two other parameters, the intensity B of an homogeneous background illumination and the dark adaptation time t_d which elapsed after turning off the room lights have also been varied. Under these conditions the response, recorded extracellularly from the terminal arborisation of the axon of a class III (ON-OFF) cell in the optic tectum, consists of an initial high-frequency burst of spikes ('early response') followed by a silent period and a long-lasting low-frequency discharge. The duration T of the silent period is the longer, the higher the flash intensity I, the lower the intensity B of the background illumination and the longer the dark adaptation time t_d. Depending upon these parameters the silent period can be as long as 30 seconds; the subsequent low-frequency discharge can last for up to another 30 seconds. The cell may, therefore, fire as late as about one minute after the flash. Similar responses have been recorded also from the terminal arborisation of other cell types and from the cell bodies in the retina (Pickering and Varjú, 1969). The experiments reported

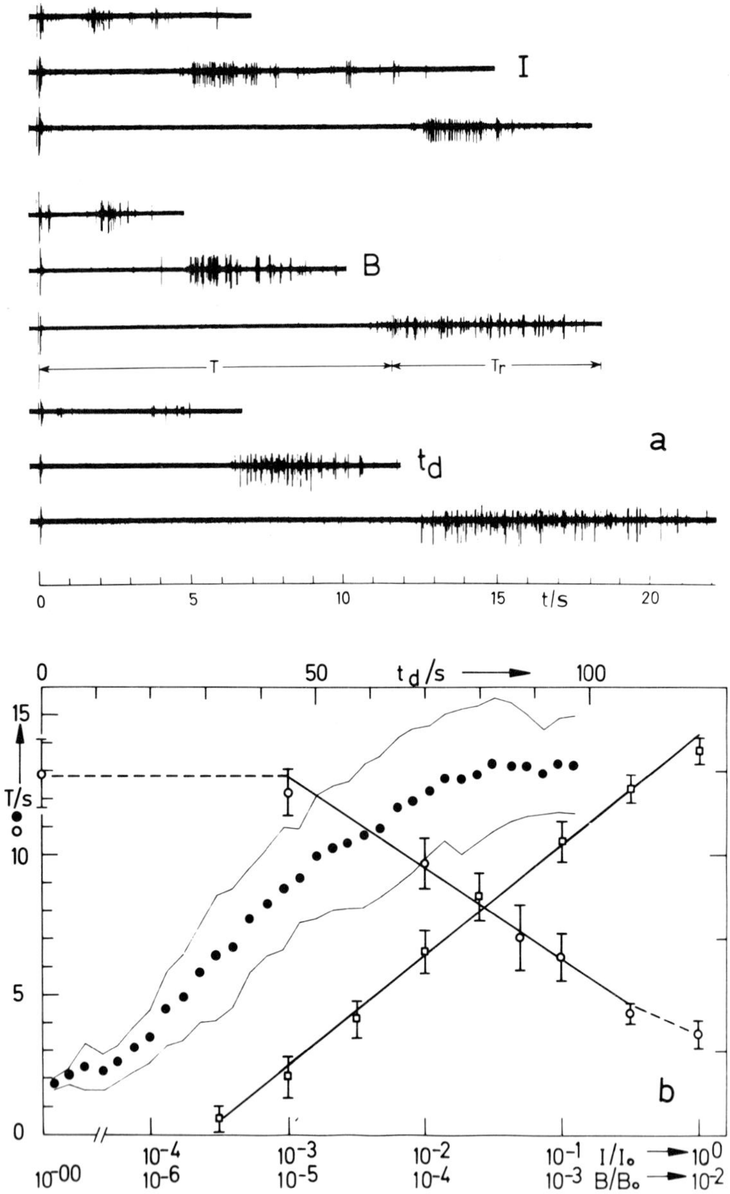

Fig. 1. Spike trains illustrating the influence of the flash intensity I, the intensity B of the background illumination (both in relative units) and the time in darkness t_d upon the duration T of the silent period and the duration T_r of the late response (a). The quantitative relations between T and the stimulus parameters I (□), B (o) as well as t_d (●, b). Bars resp. thin lines: standard error of the means. From Varjú and Pickering (1969), Pickering and Varjú (1969), modified.

below have been performed in order to shed some light upon the questions, how retinal elements respond for such a long time to short flashes and how the long silent period preceeding the low-frequency discharge can be understood. Since intracellular microprobing of ganglion cells and recording from other retinal elements are difficult in frog under the necessery experimental conditions, efforts have been made to find the answers by indirect evidence in extracellular recordings of spike activity of ganglion cells elicited by suitable stimuli. Even in extracellular recordings it is very tedious to collect experimental data, because in many cases the contact between electrode and cell has to be maintained for at least an hour, a task which can be performed only in a few percentages of all cases. Therefore working hypotheses are needed to economize the experiments. A hypothesis was offered by Sickel (1972); he suggested on the basis of recordings in the receptor layer that the response of the receptor is saturated during the silent period, and the subsequent retinal elements are exited only during the fast rising phase of the receptor response (early response of the ganglion cells) and during its slow decay (late response). Assuming proportionality between the time derivative of the receptor response and the excitation of ganglion cell, even the frequency of the ganglion cell discharge could be understood on the basis of this hypotheses. Yet it cannot be valid, because the ganglion cells can be exited also during the silent period, as will be discussed later. Pickering and Varjú (1969) postulated therefore two processes elicited by the flash, a slow excitatory (E) and a faster inhibitory (H) one,

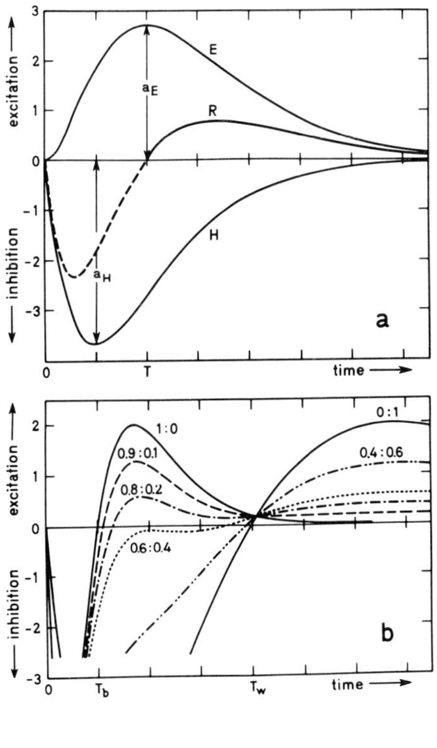

Fig. 2. Illustration of a possible mechanism capable of giving rise to the delayed response (a), and to the response obtained in experiments with inhomogeneous stimulus fields (b). From Varjú and Pickering (1969), modified.

and assumed that the discharge rate during the delayed response is a monotonously increasing function of the suprathreshold part of the sum (R) of these two processes, as illustrated in fig. 2a. The proposal was considered to be an illustration of another possible mechanism capable of giving rise to responses preceeded by an extremely long silent period rather than a quantitative model. The problems raised by this proposal were, what parameters of the curves E and H are influenced by the stimulus parameters I, B and the time t_d, and what might be the true time course of the subthreshold portion of the curve R. The inhibitory portion of the curve cannot be established directly in experiments with extracellular recordings.

In spite of these difficulties results have been presented during the last years which allow for a deeper insight into the mechanisms responsible for the delayed response (Pickering and Varjú, 1969, 1971; Varjú and Pickering 1969; Chino and Sturr 1975a, b; Heer, 1975; Steinmetz 1975). Particular attention has been paid to the dependence of the duration of the silent period upon the stimulus parameters, because it is well reproducible in subsequent series of measurements. The quantitative properties of the late discharge, i.e. the spike frequency, the total number of spikes or the duration of the response are, on the contrary, subject to strong statistical variations even in repeated experiments with the same cell. In the present paper the most important experimental findings shall be summarized and some of the consequences discussed. The details of experimental conditions, described at length in the oroginal papers, will be omitted here. All the experimental results presented have been recorded from class III neurons.

2. EXPERIMENTAL RESULTS

2.1. The time course of the ganglion cell sensitivity during the silent period

In double-flash experiments Pickering and Varjú (1969) observed that a second flash during the silent period usually elicites an early response (see fig. 4a, inset). While this finding contradicts the assumption of saturation during the silent period, it allows in connection with other experimental results to investigate the time course of the postulated inhibitory process. The additional observation is that the latency of the early response to a flash decreases with increasing flash intensity (Pickering and Varjú, 1969). It is supposed that the latency of the early response - in case of constant intensity flashes -

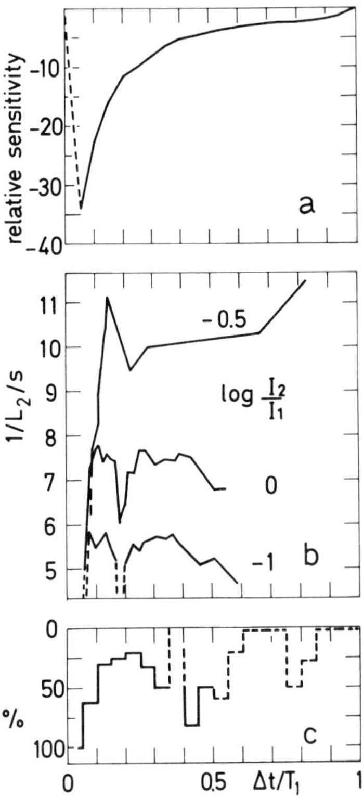

Fig. 3. Double flash stimuli. The relative sensitivity (a, average data obtained by Chino and Sturr, 1975), resp. the inverse of latency L_2 of the second early response (b, from Steinmetz, 1975, modified) as a function of the time shift Δt between first and second flash. The percentage n_0 of those cases, in which the second flash did not elicite an early response (c, from Heer, 1975). Δt is measured in units of the delay time T_1 of the late response elicited by the first flash alone. In b I_1 denotes the intensity of the first, I_2 that of the second flash. The results in b and c were obtained from - different - single cells.

is a measure of the sensitivity of the retina, which is reduced by inhibitory influences. Projecting a flash at time $t = 0$ and a second one later at $t = \Delta t$ during the silent period following the first flash, the latency L_2 of the early response to this second flash should be a monotoneously increasing function of the strength of inhibition at time $t = \Delta t$. Data have been provided by Chino and Sturr (1975) and by Steinmetz (1975). Their results are reproduced partly in fig. 3. The sensitivity function reproduced in fig. 3a (Chino and Sturr, 1975) is an average one and has been derived from the measured values L_2 using a latency-intensity characteristic. Steinmetz (data in fig. 3b) considered the inverse of L_2 as sensitivity. In both diagrams Δt is measured in units of the delay time T_1 of the late response without the second flash. Chino and Sturr (1975) obtained curves with one minimum only; they state that all individual curves become congruent after normalizing the Δt-axis, regardless of the absolute length of delay time. Steinmetz consistently obtained also a second minimum which occured at about $\Delta t/T_1 = 0.2$, again regardless of the value of T_1. Observing that the second flash does not always elicite an early response, Heer found a second way to investigate the time course of the postulated inhibition. Assuming that the probability of occurance of an early response is inversely related to the strength of inhibition, one might establish the qualitative properties of the subthreshold portion of the R-curve in fig. 2a by measuring the percentage n_o of those cases, in which no second response occures. Heer's result is shown in fig. 5c - for one pair of the intensity ratio I_2/I_1 only, - I_1 being the intensity of the first, I_2 that of the second flash. These results unequivocally document that the suggestion of Pickering and Varjú (1969) corresponds to reality: The ganglion cell is inhibited during the silent period and the strength of the inhibition changes with time. The exact time course has still to be considered as not precisely known.

2.2. The duration of the silent period in double-flash experiments

In double-flash experiments, the delay time T_2 of the late response following the second flash depends upon the time shift Δt between the two flashes. If T_2 and Δt are measured in units of T_1, latter being as defined before, again all the values of T_2/T_1 plotted over $\Delta t/T_1$ result in congruent curves, regardless of the absolute value of T_1. This is demonstrated for pairs of flashes of equal intensities in fig. 4a. In this diagram the open circles and triangles were obtained in two experiments with different values of the stimulus parameters I, B and t_d. The dots represent the average values (within sampling intervals, the length of which varies along the abscissa) obtained in many runs with widely varying values of I, B and t_d.

In addition, the data plotted in fig. 4 show how the first flash influences T_2. As long as the ratio I_2/I_1 of the flash intensities is not lower than about 1 : 10, T_2 will be increased for small values of Δt and decreased if Δt becomes larger, relative to the value T_{2s} obtained with a single flash of intensity I_2. At still higher values of Δt the first flash, of course, does not influence T_2 at all. At lower values of the ratio I_2/I_1 the delay time T_2 never becomes smaller than T_{2s}.

Finally, the third trace in the inset demonstrates again that a flash causes, - besides the early response, - a period of inhibition, because the second flash abolishes the delayed response elicited by the first flash.

2.3. The OFF-response at low stimulus intensities

In ON-OFF-cells both positive and negative steps of the stimulus intensity causes ganglion cell discharge, i.e. positive responses. The OFF-

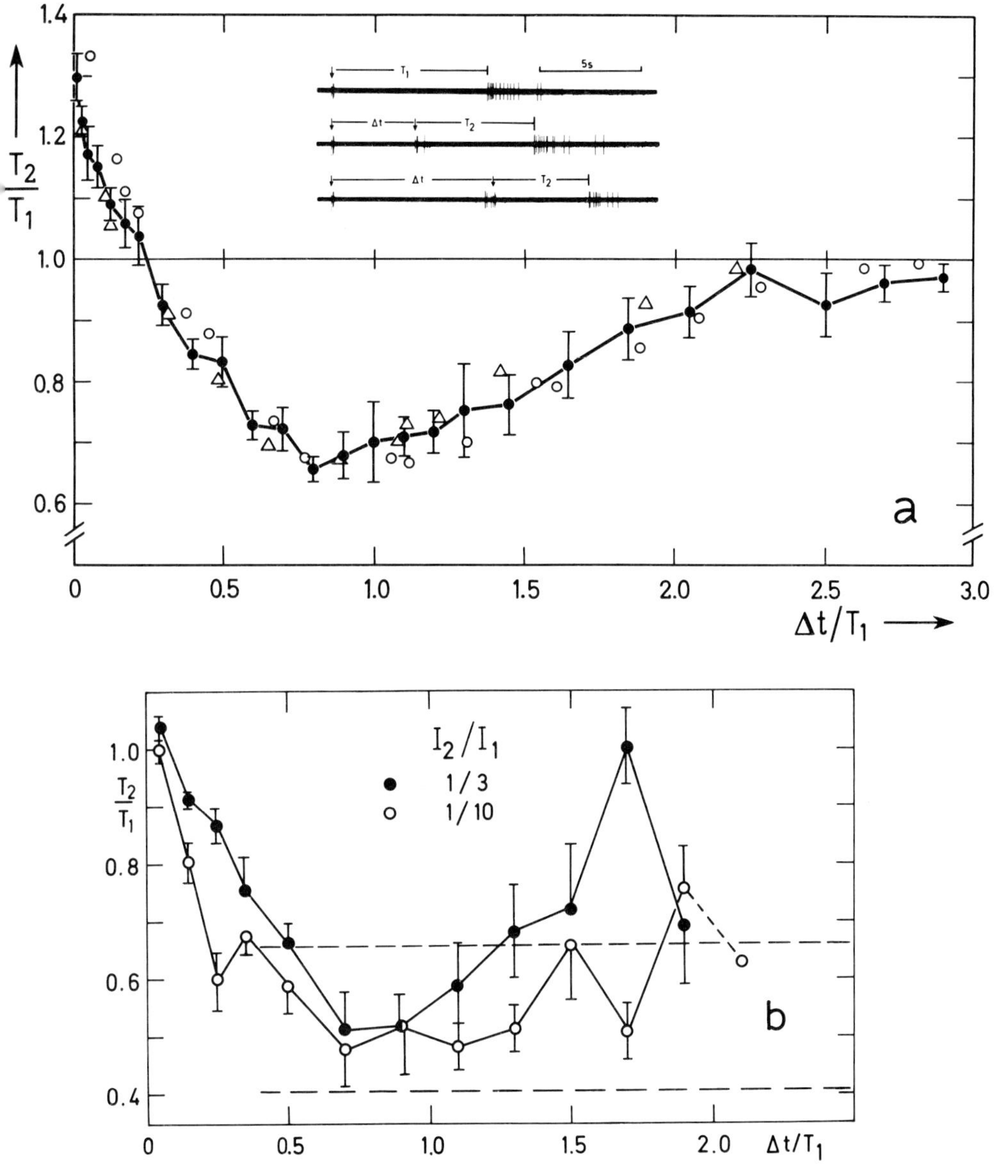

Fig. 4. Double-flash experiments. The ratio T_2/T_1 plotted over $\Delta t/T_1$; T_2 is the delay time of the late response following the second flash, T_1 the delay time of the late response if the first flash is given alone. The horizontal line in a and the broken lines in b represent the - normalized - value of T_2, if the second flash is given alone. In b the parameter is the ratio of flash intensities (which is 1 in a). Inset in a: response to a single flash (upper trace) and to double flashes with $\Delta t < T_1$ (middle trace) and $\Delta t > T_1$ (lower trace). Data in a from Varjú and Pickering (1969), in b from Steinmetz (1975), inset from Pickering and Varjú (1969).

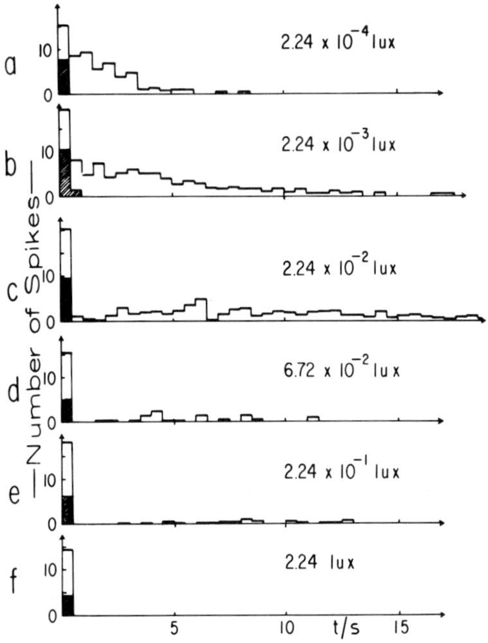

Fig. 5. Post stimulus pulse rate histograms. The shaded portions represent the cell's response to light ON, and the unshaded portions are the responses to light OFF. The order of stimulus presentation was first light OFF and then light ON. The given levels of illumination were measured in the plane of the eyes by means of a luxmeter. For all levels of illumination higher than given in f the responses were similar to that in f. From Pickering and Varjú (1971).

response may last longer and consists of more spikes than the ON-response. This is especially true at low intensity stimuli. Moreover, the duration of the OFF-response depends upon the stimulus intensity, while that of the ON-response hardly does, as demonstrated in fig. 5. The poststimulus histograms in this figure also reveal properties of the OFF-response, which closely relate it to the delayed response elicited by a flash. When the illumination amounts to 2.24 lux or even higher values - as measured in the plane of the eye - the OFF-response is a short high-frequency burst similar to the ON-response and to the early response to a flash. At progressively lower levels of illumination the initial high-frequency burst remains the same, but - gradually - a "tail" of low-frequency discharge develops. In the traces e and d of fig. 5 the tail is separated by a gap from the early, high-frequency part of the response; at still lower stimulus intensities spikes fill the gap and the response becomes shorter. Note that the entire response may - in the most favorable case - last as long as about 20 seconds, a value comparable with the duration of the delayed response elicited by a bright flash. Apparently, also OFF-stimuli can elicit "delayed" responses, which are suppressed, if the retina is pre-

Fig. 6. The interaction of step stimuli with the delayed response elicited by a flash (a, c). The influence of a preceeding ON-stimulus upon the response to a following OFF-stimulus (upper traces in b) and vice versa (lower trace in b). In the two uppermost traces in b only parts of the late responses are shown. a from Pickering and Varjú (1971), b and c from Heer (1975).

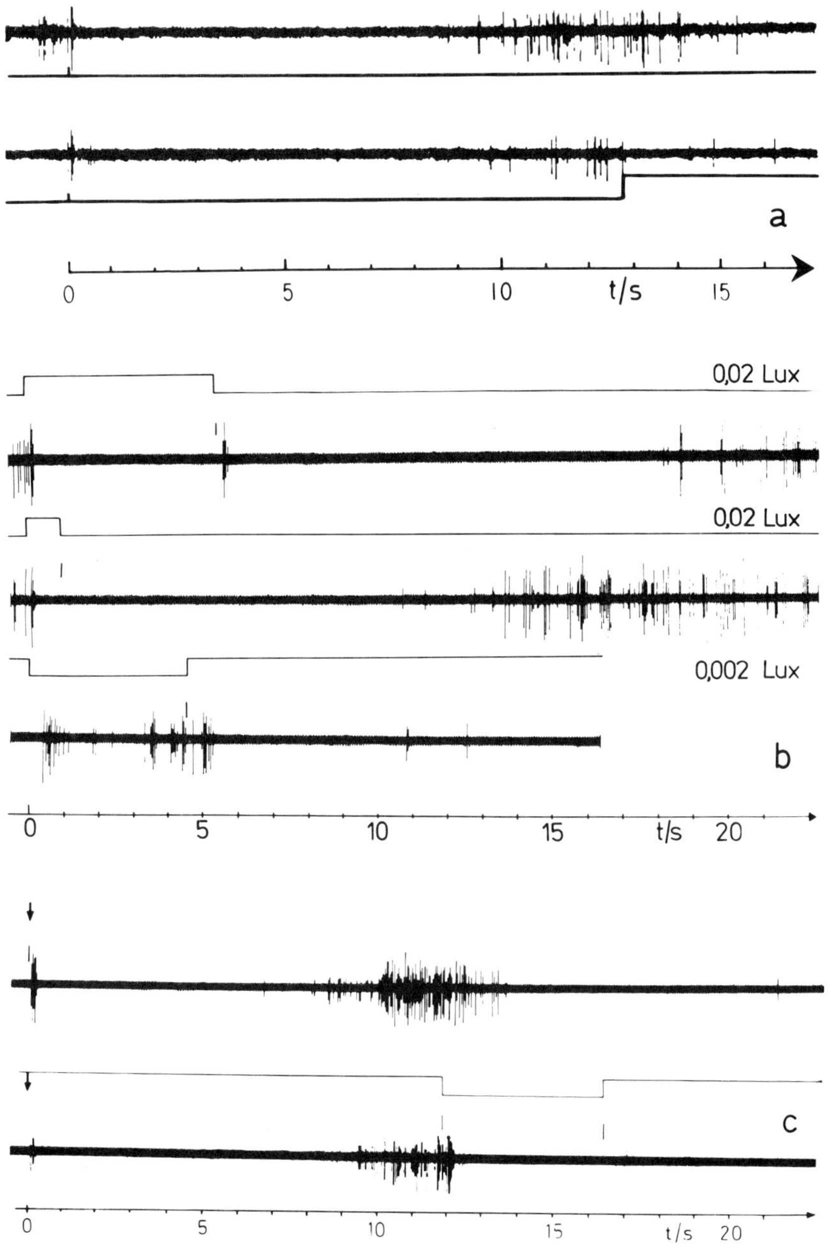

a
0
5
10
t/s
15
0,02 Lux
0,02 Lux
0,002 Lux
b
0
5
10
15
t/s
20
c
0
5
10
15
t/s
20

Fig. 6

adapted to relatively high light intensities. This result suggests that the decreasing phase of the stimulus might be responsible for the low-frequency discharge of the delayed response elicited by a flash, since it is an intensive OFF-stimulus without a strong preadaptation.

2.4. Interaction of ON- and OFF-stimuli with the delayed response

Pickering and Varjú (1971) observed that an ON-stimulus suppresses the delayed response just as well as a second flash. They concluded that the rising phase of a flash might be responsible for the postulated inhibitory process H (fig. 2a). One would, therefore, expect that also an OFF-response will be abolished by a preceeding ON-stimulus, if only the time shift between the ON- and OFF-stimuli is sufficiently short. Fig. 6b shows that this actually happens.

Heer (1975) demonstrated that also an OFF-stimulus suppresses the delayed response, as shown in fig. 6c. Therefore we have to conclude that also an OFF stimulus generates both excitatory and inhibitory processes, much like to the curves shown in fig. 2a. The inhibitory process seems to be weak at low intensities and to increase faster than the excitatory process, when the intensity of the preadaptation is increased.

2.5. Stimulation of restricted areas of the receptive field (RF)

Since responses of retinal ganglion cells are determined to a large extent by lateral interactions within the RF, the question arises whether such interactions influence also the delayed response, in particular the duration T of the silent period. In an experiment, the results of which are shown in fig. 7, Heer (1975) stimulated restricted areas within the RF. The stimulated area was a vertical stripe of varying width, the stimulus a short flash.

If the angular width $\Delta\alpha$ of the illuminated area is reduced, T remains unchanged down to $\Delta\alpha = 1.7^{\circ}$. Below this value T decreases rapidly with decreasing $\Delta\alpha$, but it is uncertain, whether the extent of the stimulated area is decisive for the value of T. It is possible that the width of the stimulated area cannot be reduced much below the critical value of $\Delta\alpha = 1.7$ in the paralized frog due to the dioptric properties of the eye. Further reducing the width of the stripe would result only in a decrease of the stimulus intensity and, therefore, of T. Since the effect of lateral interactions within the RF can usually be altered by changing the size or the location of the stimulated area, we

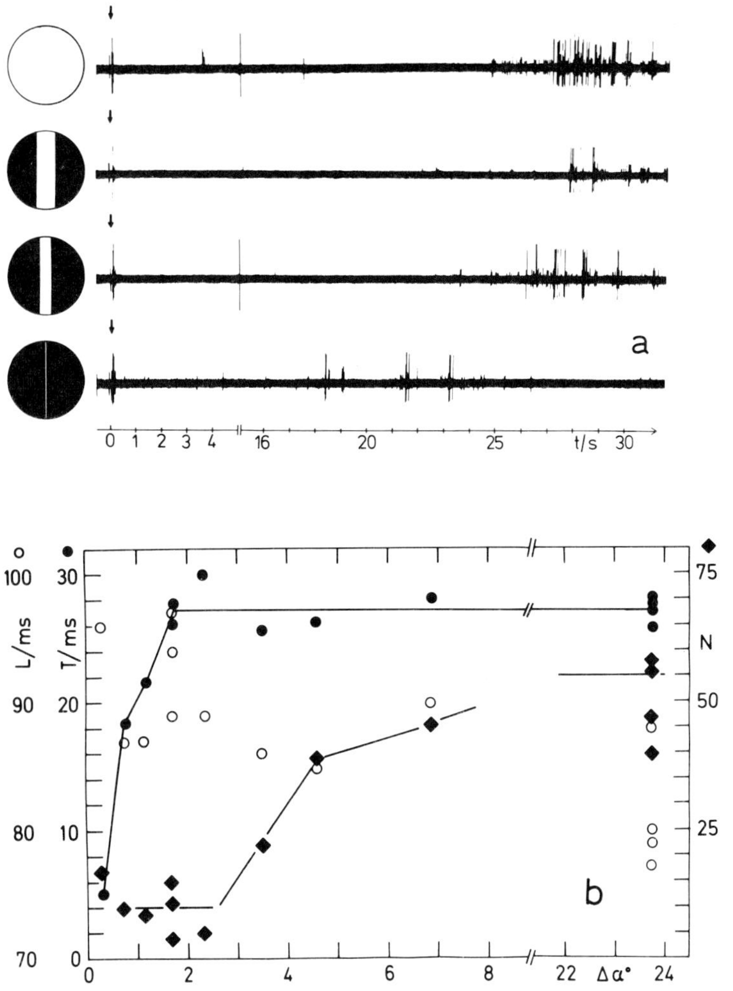

Fig. 7. Spike trains elicited by flashes projected into restricted areas of the RF (a) and the dependence of the duration T (●) of the silent period, the total numer of spikes N (◆) during the delayed response and the latency L (o) of the early response upon the size of the stimulated area (b). The stimulus field was a vertical stripe of varying width Δα. In the three uppermost traces in a only the beginning of the late response is shown. From Heer (1975), modified.

might expect T not to depend upon lateral interactions. Even if there is some influence of such an interaction, its range is very narrow compared with the size of the RF, the diameter of which in class III cells being usually larger than 12°.

2.6. Inhomogeneous stimulus fields

Further support for the independence of T from lateral interactions yielded experiments in which different areas of the RF have been stimulated with flashes of different intensity. Fig. 8 shows again results obtained by Heer (1975). He subdivided the RF along a vertical straight line and illuminated the two parts homogeneously with flashes of different intensity selecting three different contrast ratios. If the contour line was in one of the extrem positions, the entire RF was stimulated either with the brighter or the darker flash alone, and the delayed discharge began after the corresponding delay times T_b and T_d, respectively, T_b being larger than T_d. Yet when the RF was subdivided by the contour line in the angular position $\alpha = 0^o$ into two - presumably equal - parts, two distinct delayed responses appeared with delay times about T_b and T_d.

The two responses are the better separated, the higher the contrast ratio. If the contour line is moved away from the $\alpha = 0^o$ position, the number of spikes in one of the responses decreases and diminuishes, in the other one increases, while the values of the delay times T_d and T_b do not change appreciably (fig. 8b, c). A similar experiment with similar result has been carried out earlier by Varjú and Pickering (1969) on one cell and only with one contrast ratio. They subdivided the RF into two parts along a circle centered upon the RF.

Would the total light flux impinging upon the RF of the ganglion cell determine the value of T, one should expect T to change gradually from T_d to T_b (or vice versa), as the contour line moves from one extreme position towards the other. The results clearly exclude spatial integration of stimulus intensity over the RF. Yet the outcome of the experiments can easily be interpreted assuming parts of the RF to contri-

Fig. 8. Inhomogeneous stimulus fields. Poststimulus histograms (a, number of spikes per 0.25 s, delayed responses only). The dependence of the total number of spikes N_d, N_b of the delayed response (b) as well as the delay times T_d and T_b (c) upon the angular position α of the contour line subdividing the RF into adjacent areas stimulated with dim and bright flashes. Symbol identification in c holds also for b. From Heer (1975).

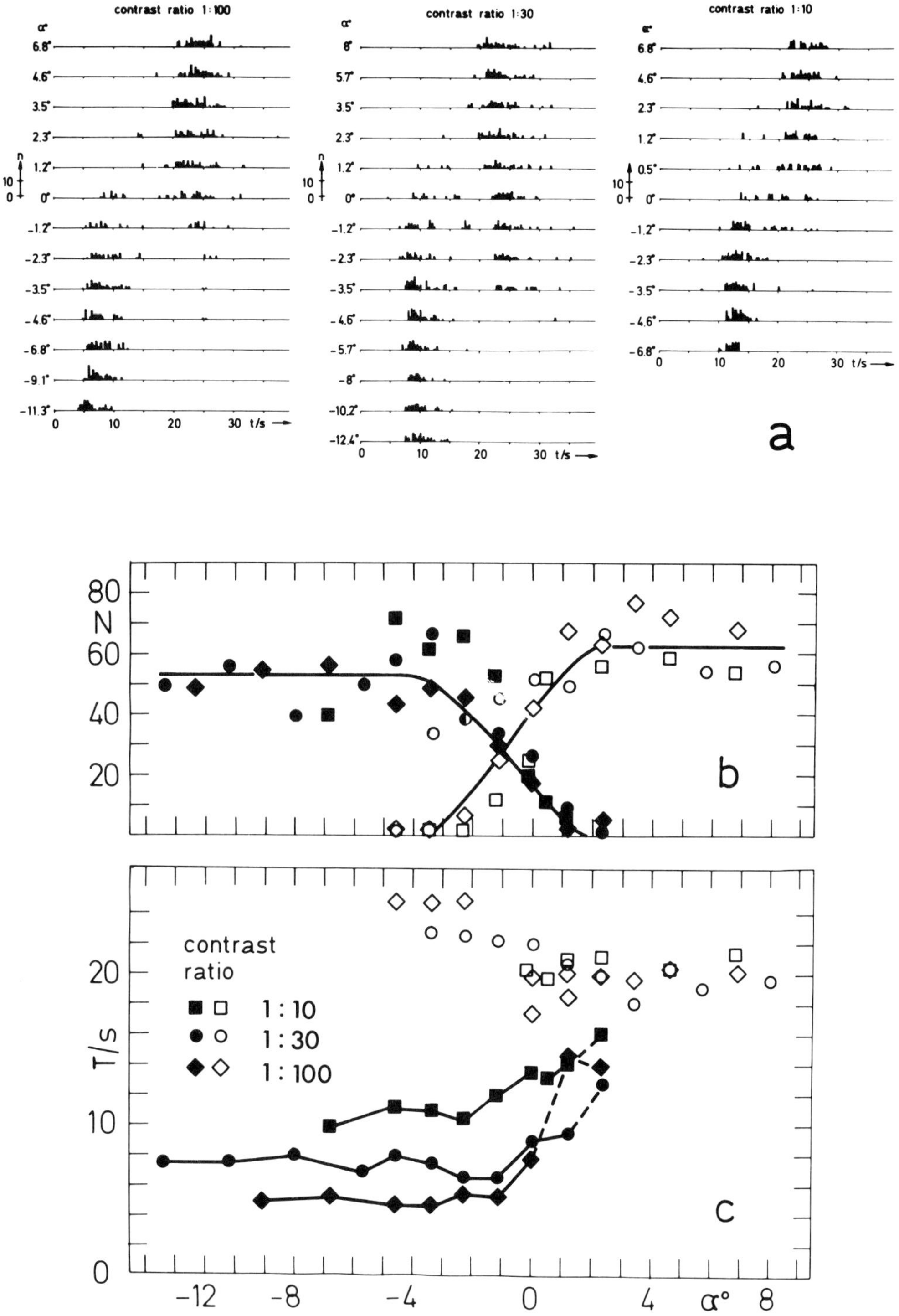
contrast ratio 1:100
contrast ratio 1:30
contrast ratio 1:10
t/s
a
N
b
T/s
contrast ratio
1 : 10
1 : 30
1 : 100
c
α°

Fig. 8

bute independently to the excitation of the cell, the time course of the contribution being determined by the local flash intensity. Instead of the stimulus, partial responses should be integrated at the ganglion cell. In the present case the two partial responses would - according to this assumption - be the stronger the larger the corresponding portion of the RF is. Now, if - say, the brighter - portion of the RF increases starting from zero, the corresponding and initially subthreshold excitation becomes stronger and stronger, reaching first just suprathreshold and finally full strength. The excitation generated in the darker part of the RF behaves in the opposite way.

This would also explain the slight change of T_d and T_b as well as the gradual increase respectively decrease of the total number of spikes in the transition range, as illustrated in fig. 2b. There, the relative strength of two arbitrarily chosen R-curves has been varied as indicated by the given ratios. The zero-crossing, or may be the crossing of a higher non-zero threshold of the sum of the two curves determine the values of T_d and T_b, while the area between the curves and the threshold line corresponds to the total number of spikes.

The consequence of these results is: no far-reaching spatial interactions have to be taken into account in trying to find the origin of the delayed response.

2.7. The variability of ganglion cell response

The qualitative and quantitative details of a ganglion cell response can appreciably deviate from those of other cells of the same type. As already mentioned in chapter 2.1. the second flash given during the silent period does not always elicites an early response. Moreover, the total number of spikes and the duration of the delayed response are subject to strong variations. Furthermore, it was observed during the course of experimentation that a preceeding ON-stimulus can fail to supress a following OFF-response, even if the time shift between the two stimuli is short enough (c.f. chapter 2.4. and see Heer, 1975). In some cells the early response consists of two well separated bursts of spikes (see e.g. fig. 7 in Varjú and Pickering, 1969). Such responses can particularly often be observed after the second flash in double-flash experiments, if the intensity I_2 of the second flash is lower than that of the first one, I_1. In those cases the time shift T_z between flash and the second group of spikes increases with increasing time shift Δt between the two flashes (Steinmetz, 1975). Here again,

T_z/T_1 plotted over $\Delta t/T_1$ results in congruent curves, no matter what the stimulus conditions are (c.f. chapters 2.1. and 2.2.).

Even if some or possibly most of these variations of the ganglion cell response are due to different physiological states of the preparation, they have some implications and consequences. First, some of the discrepancies between results obtained in different laboratories, like those between the results of Chino and Sturr (1975 a, b) on the one hand and of Steinmetz (1975) and Heer (1975) on the other hand (c.f. chapter 2.1.) may be consequences of this variability of the response. Secondly, any hypothesis on the ganglion cell response should be capable to account for this variability, if only some plausible variations of the parameters can be assumed.

3. MODEL CONSIDERATION AND DISCUSSION

The block diagram in fig. 9 is an attempt to summarize the consequences we can draw from the discussed experimental results. Since step stimuli do not cause sustained responses, a first order high pass filter with time constant τ is assumed to act upon the signal generated by the stimulus. The diodes symbolize rectification and separate, thus, the ON (w_{Γ}) and OFF ($w_{\urcorner}$) components of the response, which are further processed separately. Since both ON and OFF components have been shown to cause excitatory as well as inhibitory influences, a further branching of both ON and OFF channels into excitatory (E_{Γ}, $E_{\urcorner}$) and inhibitory (H_{Γ}, $H_{\urcorner}$) paths is postulated. A sign change in the - lower - OFF-branch accounts for the positive response to the OFF stimulus. The assumption that the ON and OFF components are processed sepa-

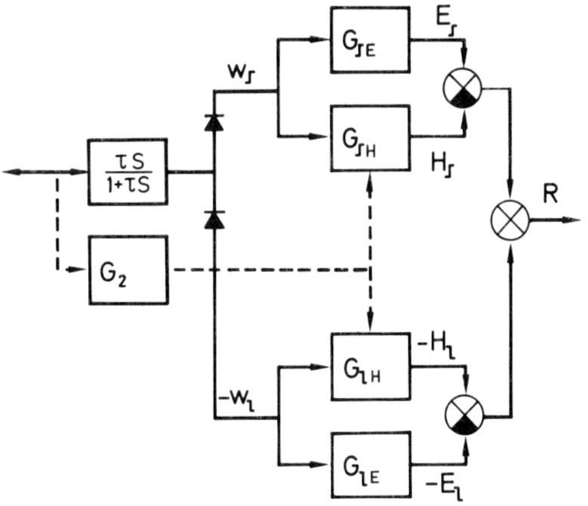

Fig. 9. Block diagram summarising the experimental findings and the consequences drawn. The symbols of diods represent rectification, the sectored circles addition. The black sectors indicate change of sign. For further explanation see text.

rate channels is strongly supported by recent experimental results obtained by Levine and Shefner (1977) in the fish retina. Separate paths for excitatory and inhibitory influences are very likely, since these are mediated as a rule by different synapses. Whether the paths converge upon the ganglion cell or prior to it is irrelevant regarding the model. Additivity of the outputs of the four channels, as indicated, is the most simple attempt to describe their integrated action upon the ganglion cell.

The task is to find proper transfer operators $G_{\ulcorner E}$, $G_{\ulcorner H}$, $G_{\urcorner E}$, $G_{\urcorner H}$, which describe quantitatively the experimental results. One difficult question is, how to account for the influence of background light upon the delayed response. Since an increase of the background illumination has the same effect upon the duration T of the silent period as a decrease of the flash intensity, it could act prior to the high pass filter by reducing the effectivness of the flash stimulus. Assuming further that this influence declines very slowly after turning off the background light, even the dependence of T upon the dark adaptation time t_d could be accounted for in this way. The transfer properties of the four channels should, then, depend also upon the input pulses delivered by the high pass filter. As shown in chapters 2.1 and 2.2, response curves obtained in experiments with different stimulus parameters become congruent, if only the time is normalized, i.e. measured in units of the delay time T resp. T_1. This suggest that the change of the delay time T as a function of the variables I, B and t_d might be the consequence of a corresponding change of the time scale; such a change of the time scale occures e.g. if all the relevant time constants will be altered by the same factor. Thus the input to the four channels should influence the response in such a way that the time scale is changed by the effective stimulus according to the power function marked by open squares in fig. 1b.

Nevertheless, the possibility has also to be considered that a sustained response reaches - parallel to the transient and rectified signals - the transmitting stages in the channels $E_{\ulcorner}$, $H_{\ulcorner}$, $E_{\urcorner}$, $H_{\urcorner}$ and modifies their parameters. This is indicated - only for the inhibiting paths - by dashes connections.

Analog computer simulation, although not yet completed, revealed several basic properties of the model. Linear filters of low pass type realized both inhibitory and excitatory paths. The time constants of the filters and the relative weight of the channels have been changed in discrete steps rather than continuously under the influence of the

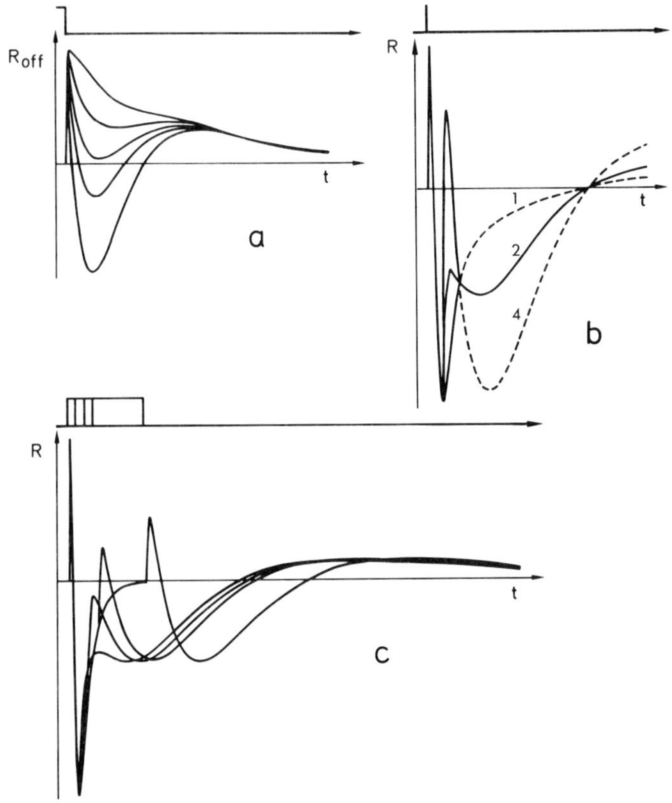

Fig. 10. Analog computer simulation of the model in fig. 9. Sample registration of responses. For further details see text.

stimulus. Fig. 10 shows sample registrations. The suprathreshold parts of the curves (above the time axis) represent the frequency of ganglion cell discharge, the subthreshold parts (below the time axis) the net inhibition. In fig. 10a the influence of increasing stimulus intensity - from top to bottom - upon the OFF-response is simulated. The curves resemble very closly the post stimulus histrograms in fig. 5. In fig. 10b puls responses are shown. Note the qualitative change of the net inhibitory curves, as the relative weight of the response to the decreasing phase of the puls is varied as indicated. A similar variation of the contribution of the response elicited by the decreasing phase might account for the differences between the curves in fig. 3 as well as for the second early response observed occasionally. Finally, in fig. 10c the responses to pulses of different duration are simulated. In case of short pulses the OFF-response is suppressed by the ON-inhibition, like in fig. 6b.

While these and several other qualitative properties of the ganglion cell response can easily be reproduced by the model, the quantitative simulation of the results obtained in e.g. double-flash experiments is at present less satisfactory. Nevertheless it is expected that continuation of these model studies will contribute substantially to the solution of the problem.

In connection with the delayed response usually the question is asked what might be its physiological significance? According to behavioral studies of Ingle (1971) it is presumably the physiological basis of afterimages. Yet even without physiological significance it should not be omitted while studying neural mechanisms in the retina: ignoring this strange response one might easily overlook some properties of the system, which are essential with respect to both neural mechanisms and behaviour.

Supported by the Deutsche Forschungsgemeinschaft, grant No. Va 22/9

1) Barlow, H.B.: Action potentials from the frog's retina. J. Physiol. 119, 58-68 (1953)
2) - Summation and inhibition in the frog's retina. J. Physiol. 119, 69-88 (1953)
3) Chino, Y.M. and Sturr, J.F.: The time course of inhibition during the delayed response of the ON-OFF ganglion cell in the frog. Vis. Res. 15, 185-191 (1975 a)
4) - Rod and cone contributions to the delayed response of the ON-OFF ganglion cell in the frog. Vis. Res. 15, 193-202 (1975 b)
5) Grüsser, O.-J., Grüsser-Cornehls, U.: Neurophysiology of the anuran visual system. In: Frog Neurobiology. Llinás, R., Precht, W. eds. Springer, Berlin - Heidelberg - New York, 1976
6) Hartline, H.K.: The response of single optic nerve fibers of the vertebrate eye to illumination of the retina. Am. J. of Physiol. 121, 400-415 (1938)
7) - The receptive fields of optic nerve fibers. Am J. of Physiol. 130, 690-711 (1940)
8) Heer, I.: Zeitliche und räumliche Wechselwirkungsprozesse bei der Entstehung der Spätantwort retinaler Ganglinezellen im Frosch. Dissertation, Universität Tübingen, 1975
9) Ingle, D.: A possible behavioural correlate of delayed retinal discharge in anurans. Vis. Res. 11, 167-168 (1971)
10) Levine, M.W., Shefner, I.M.: Variability in ganglion cell firing patterns: Implication for separate "ON" and "OFF" processes. Vis. Res. 17, 765-776 (1977)
11) Maturana, H.M., Lettvin, J.Y., MCCulloch, W.S., Pitts, W.H.: Anatomy and physiology of vision in the frog (Rana pipiens). J. gen. Physiol. 43, 129-175 (1960)

12) Pickering, S.G. and Varjú, D.: Ganglion cells in the frog retina: inhibitory receptive field and long-latency response. Nature 215, 545-546 (1967)
13) - Delayed response of ganglion cells in the frog retina: the influence of stimulus parameters upon the length of the delay time. Vis. Res. 9, 865-879 (1969)
14) - The retinal ON-OFF components giving rise to the delayed response. Kybernetik 8, 145-159 (1971)
15) Sickel, W.: Retinal metabolism in dark and light. Handbook of Sensory Physiol. Vol. VII/2, 667-727 (1972)
16) Steinmetz, R.: Untersuchungen zur Entstehung der Spätantwort retinaler Ganglienzellen im Frosch. Dissertation, Universität Tübingen, 1975
17) Varjú, D. and Pickering, S.G.: Delayed response of ganglion cells in the frog retina: the influence of stimulus parameters upon the discharge pattern. Kybernetik 6, 112-119 (1969)
18) Varjú, D.: Functional classification and receptive field organization of retinal ganglion cells in the frog. In: Processing of optical data by organisms and mashines. W. Reichardt, ed., Academic Press, London, 1969

ON CONSTRAINED TRANSINFORMATION FUNCTIONS*

Roland Heim
Institute for Information Sciences
University of Tübingen
Fed. Rep. of Germany

Introduction

The classical, statistically founded information theory is a useful tool in describing how data are processed by man-made systems. But a comparably small progress has been made in the application of information theory to biological systems and their data processing. There are two main reasons for this fact.

The first one is that a purely statistical definition of information is too narrow to cover all phenomena with inherent structures and regularities. A very promising extension to overcome those limitation is the concept of program complexity first introduced in Kolmogoroff (1964). The measure of information content is the compressibility of a data set by means of a Turing machine. It is possible to build up an algorithmical information theory which contains the statistical theory conceptually as a special case. See e.g. Heim (1976a, 1976b).

The other reason is that the classical theory was mainly concerned with the lossless reconstruction of encoded data resp. with the possibility of an asymptotically lossless transmission of information. But for a fruitful application of information theory one has to account for the large data suppression in biological infosystems as a permanent decision process between relevant and irrelevant information. Considering the huge amount of information in our environment, data suppression is a necessity for any real system. On the other hand, the survival of a biological infosystem requires a minimum amount of specifically selected information. From a simple point of view, the information processing structure of such an infosystem reflects an equilibrium between to conflicting tendencies: To keep the processing capacity high in order to survive and to keep it low for processing capacity is expensive.

Adopting the classical source - channel - receiver scheme, a natural way to quantify additional qualitative properties of data and events is to assign numerical values to the source (channel input) pro-

* Supported by the Deutsche Forschungsgemeinschaft
The paper is dedicated to the memory of Ernst Pfaffelhuber

babilities resp. to the channel transition probabilities as the determining quantities of the transinformation. Depending on the interpretation, any value of transinformation is associated with an average cost, utility, distortion etc. For instance, if every input of a channel has a specific cost (electric power, transmission time), a typical problem is how to choose the input probabilities to have maximal transinformation for a prescribed maximal average cost. Conversely, if the source is fixed and some imprecision in its reproduction is tolerable, one may look for the lowest transinformation by varying the transition probabilities such that a prescribed distortion is not exceeded.

In spite of the severe limitations of the statistical theory, the constrained transinformation exhibits some interesting properties of great practical significance. For instance, even a small amount of tolerable imprecision saves a considerable amount of channel capacity. Or to gain the last percents of the channel capacity requires a disproportionate amount of costs as defined above. Observations of this kind are well known in the everyday life, e.g. the learning of a foreign language.

On the other hand, the advantage of such simple models is that they are within the range of analytical techniques. Though not all solutions are calculable analytically, it is nevertheless possible to make general statements on their properties.

The following discussion deals only with finite, discrete alphabeths which is sufficient to show the essential features of the constrained transinformation functions. It seems that only Blahut (1972) outlined a formalization of the constrained capacity function whose basic properties are presented in the first chapter in some detail. The next chapter restates the fundamentals of the rate-distortion theory as it may be found in Gallager (1968) or Berger (1972), but with some new proofs. Chapter 3 presents some analytically treatable examples and its implications, e.g. an information theoretic interpretation of the existence of feature detectors in the sensory pathway. The results in Pfaffelhuber (1974) on minimum cost transmission of information are special cases of the capacity-expense function.

1. The Capacity-Expense Function

The simplest model of a constrained channel is a discrete channel defined by a transition matrix $\mathbb{Q} = \{Q_{kj}\}_{j = 1...L, k = 1...M}$, $L,M \geq 2$, where an expense e_j is associated to each input letter j. The problem is to find the maximum of the transinformation $T(p,\mathbb{Q})$, the constrained channel capacity, such that the average expense is not more than a prescribed real number E. Let p (q) be the vector of input (output) probabilities of the given channel with components $p_j, j = 1...L$ (q_k, $k = 1...M$). Then we have the following convex optimization problem:

$$C(E,\mathbb{Q}) = \max_p T(p,\mathbb{Q}) = \max_p \sum_{j,k} p_j Q_{kj} \log \frac{Q_{kj}}{q_k} \tag{1.1a}$$

$$\sum_j p_j e_j \leq E \tag{1.1b}$$

$$\sum_j p_j = 1 \tag{1.1c}$$

$$p_j \geq 0 \ , \ j = 1...L \tag{1.1d}$$

Since the set $P_E = \{p | e^t p \leq E\}$ is contained in the set $P_{E'}$ if and only if $E \leq E'$, $C(E,\mathbb{Q})$ as a function of E is nondecreasing and the solution lies always on the boundary of the convex set P_E. Thus we can replace the constraint (1.1b) by the equality $e^t p = E$. Without the constraint (1.1d) which is the main obstacle to an analytical solution of this convex programming problem we have to determine any stationary point p^* from the Lagrange function

$$L(p,\mathbb{Q}) = T(p,\mathbb{Q}) - s\sum_j p_j e_j - (t - 1)\sum_j p_j$$

with Lagrange multipliers s,t and the requirement

$$\frac{\partial L}{\partial p_j} = \sum_k Q_{kj} \log \frac{Q_{kj}}{q_k} - se_j - t = 0 \qquad j = 1...L \tag{1.2}$$

If a solution turns out to have only nonnegative components p_j^*, then the complete problem has been solved. If not, the convexity of $T(p,\mathbb{Q})$ implies that the maximum is attained at a point p^* where one or more components p_j^* are zero: the solution lies on the boundary. We define the interior set $I = \{j | p_j > 0\}$, and the boundary set $B = \{j | p_j = 0\}$.

First we shall show that for the complete problem (1.1) the conditions (1.2) have to be replaced by

$$\sum_k Q_{kj} \log \frac{Q_{kj}}{q_k} - se_j = t \qquad j \in I$$
$$(1.3)$$
$$\sum_k Q_{kj} \log \frac{Q_{kj}}{q_k} - se_j = t_j \leq t \qquad j \in B$$

Let $\bar{p}^* = p^* + \Delta p$ be a variation of a solution vector p^* of (1.3) such that the average expense E remains unchanged. Explicitely: $\sum_j \Delta p_j = 0$, $\sum_j \Delta p_j e_j = 0$. Then from (1.3) it follows

$$\Delta T(p^*,\mathbb{Q}) = \sum_j \Delta p_j \frac{\partial T(p^*,\mathbb{Q})}{\partial p_j} = s\sum_j \Delta p_j e_j + \sum_{j\in I} \Delta p_j t + \sum_{j\in B} \Delta p_j t_j$$

$$= t \sum_{j\in I} \Delta p_j + \sum_{j\in B} \Delta p_j t_j - \sum_{j\in B} \Delta p_j t = \sum_{j\in B} \Delta p_j (t_j - t) \leq 0 \qquad (1.4)$$

The last inequality in (1.4) holds since Δp_j must be nonnegative if p_j^* is zero. Therefore, p^* yields a relative and because of the convexity of $T(p,\mathbb{Q})$ an absolute maximum. On the other hand, if p is a probability vector such that $t_j > t$, $j \in B$, then $\Delta T(p,\mathbb{Q}) > 0$ for any **admissible** variation with $\Delta p_j > 0$ and p cannot be a maximum point. QED

Multiplying both sides of (1.3) with p_j^* and summing over j gives us the parametric representation

$$C(E_s,\mathbb{Q}) = sE_s + t_s \qquad (1.5)$$

where the index s indicates the mutual coupling of s,t,E. The usual way to determine the capacity-expense function is to select a numerical value for the parameter s and then to solve (1.3) by analytical or numerical methods. See Blahut (1972) for a simple and always converging computer algorithm.

Now let us establish some basic properties of the constrained capacity function.

<u>Theorem 1.1</u>

The function $C(E,\mathbb{Q})$ is convex $\cap$ with respect to E and convex $\cup$ with respect to $\mathbb{Q}$.

<u>Proof</u>

a) Let p',p'' be probability distributions such that $C(E',\mathbb{Q}) = T(p',\mathbb{Q})$, $C(E'',\mathbb{Q}) = T(p'',\mathbb{Q})$ and p^β the convex combination $p^\beta = \beta'p' + \beta''p''$. Since $e^t p^\beta = \beta' e^t p' + \beta'' e^t p'' = \beta'E' + \beta''E''$, p^β is element of the set $P_{\beta'E'+\beta''E''}$ and we have

$$C(\beta'E'+\beta''E'',\mathbb{Q}) \geq T(p^\beta,\mathbb{Q}) \geq \beta'T(p',\mathbb{Q}) + \beta''T(p'',\mathbb{Q})$$
$$= \beta'C(E',\mathbb{Q}) + \beta''C(E'',\mathbb{Q})$$

where we have used the convexity $\cap$ of $T(p,\mathbb{Q})$ with respect to p.

b) Let p',p'',p^{β} be probability vectors such that $C(E,\mathbb{Q}') = T(p',\mathbb{Q}')$, $C(E,\mathbb{Q}'') = T(p'',\mathbb{Q}'')$, and $C(E,\mathbb{Q}^{\beta}) = T(p^{\beta},\mathbb{Q}^{\beta})$ where $\mathbb{Q}^{\beta}$ is the convex combination $\beta'\mathbb{Q}' + \beta''\mathbb{Q}''$. By definition, p',p'',p^{β} are elements of P_E. Part b) of the theorem follows from the inequalities

$$\beta'T(p',\mathbb{Q}') + \beta''T(p'',\mathbb{Q}'') \geq \beta'T(p^{\beta},\mathbb{Q}') + \beta''T(p^{\beta},\mathbb{Q}'') \geq T(p^{\beta},\mathbb{Q}^{\beta})$$

The first inequality holds since $C(E,\mathbb{Q})$ is defined as a maximum over all admissible distributions, the second uses the convexity $\cup$ of $T(p,\mathbb{Q})$ as function of the transition matrix $\mathbb{Q}$. QED

In order to simplify the following discussion, we assume any expense vector e to have only nonnegative components and at least one component to be zero. This is possible without restricting generality since we can always apply the normalization $e_j = e'_j - \min_i\{e'_i\}$ and because of the linearity of the expectation e^tp this shifts the capacity-expense function simply along the E-axis by the amount $\min_i\{e_i\}$. It is understood that all components e_j are finite.

Obviously, the maximal value of $C(E,\mathbb{Q})$ for a given transition matrix $\mathbb{Q}$ is the channel capacity $C(\mathbb{Q})$. Inspection of (1.3) reveals that the channel capacity is reached for $s = 0$. As a convex and nondecreasing function $C(E,\mathbb{Q})$ is continous and strictly increasing with respect to E. Therefore we can write

$$\lim_{E\to E_{max}} C(E,\mathbb{Q}) = C(\mathbb{Q}) \tag{1.7}$$

where $E_{max} = e^tp^o$ and $C(\mathbb{Q}) = T(p^o,\mathbb{Q})$.

Now we are going to study the behavior of the capacity-expense function when the parameter s runs through all positive reals. First we observe from (1.5) and the Lagrange function $L(p,\mathbb{Q})$ that

$$t_s = \max_p \{ T(p,\mathbb{Q}) - se^tp \} \tag{1.8}$$

The next theorem is a geometrical interpretation of (1.5).

<u>Theorem 1.2</u>

If the parameter s generates the point (E_s,C_s), then $y = sx + t_s$ is a tangent line at this point.

<u>Proof</u>

Assume there are two points (E_s,C_s), (E',C') such that

$$\begin{aligned} C_s &= sE_s + t_s \\ C' &= sE' + t_s \end{aligned} \tag{1.9}$$

Then $C' = s(E' - E_s) + C_s$. On the other hand we have by definition

$$C' = \max_{p \in P_{E_s}} \{T(p,\mathbb{Q})\} - sE_s + sE' = \max_p \{T(p,\mathbb{Q}) - se^t p\} + sE'$$

$$= \max_p \{T(p,\mathbb{Q}) - s(e^t p - E')\} \tag{1.10}$$

From (1.9) and (1.10), we see that s generates (E',C') and we have either $E' = E_s$ or s generates a straight line segment between the points (E',C') and (E_s,C_s). QED

We see from theorem 1.2 that the parameter s is the slope of the capacity-expense function at a point generated by s, that is, $s = dC/dE$ provided that C has a derivative at this point. t_s is the C-axis intercept of the tangent line at this point.

As a convex function, $C(E,\mathbb{Q})$ has a right and left derivative everywhere. Now we make precise the conditions under which left and right derivative may be different. To this end take a value $E_o < E_{max}$. There are two possibilities: 1. There are two elements i,j in the interior set I_o such that $e_i \neq e_j$. Then the parameter s is uniquely determined, otherwise we have the contradiction

$$\begin{aligned} (s' - s'')e_i &= t'' - t' \\ (s' - s'')e_j &= t'' - t' \end{aligned} \qquad s' \neq s'', \; t' \neq t'' \tag{1.11}$$

which follows from (1.3). 2. The other possibility is $e_i = e_j$ for all i,j in I_o. Since $e_j = E_o$ for all $j \in I_o$, there must be for a sufficiently small $\varepsilon > 0$ a $i \in B_o$ such that p_ε^+ is a solution vector to $E_o + \varepsilon$ and $e_i > E_o, i \in I_\varepsilon^+$, also there must be a $j \in B_o$ such that p_ε^- is a solution vector to $E_o - \varepsilon$ and $e_j < E_o$, j I_ε^-. With the abbreviation $T_j = \sum_k Q_{kj} \log Q_{kj}/q_k$ we have from (1.3)

$$\begin{aligned} T_i(p_\varepsilon^+) &= s_\varepsilon^+ e_i + t_\varepsilon^+, \quad T_i(p_\varepsilon^-) \leq s_\varepsilon^- e_i + t_\varepsilon^- \\ T_j(p_\varepsilon^-) &= s_\varepsilon^- e_j + t_\varepsilon^-, \quad T_i(p_\varepsilon^+) \leq s_\varepsilon^+ e_j + t_\varepsilon^+ \end{aligned} \tag{1.12}$$

and all parameters $s_\varepsilon^\pm$, $t_\varepsilon^\pm$ are uniquely determined since case 1. holds. The continuity of all involved functions implies that

$$\lim_{\varepsilon \to 0} T_{i,j}(p_\varepsilon^\pm) = \lim_{\varepsilon \to 0} T_{i,j}(p_\varepsilon^\mp)$$

and we have

$$s^+e_i + t^+ \leq s^-e_i + t^-$$

$$s^+E_o + t^+ = s^-E_o + t^- \qquad (1.13)$$

$$s^+e_j + t^+ \geq s^-e_j + t^-$$

where $s^+ = \inf_\varepsilon s_\varepsilon^+$, $s^- = \sup_\varepsilon s_\varepsilon^-$ and t^-, t^+ analogously. The conditions (1.3) for $s^\pm, t^\pm$ may be written as

$$\begin{aligned} T_h(p_o) &= s^+E_o + t^+ = s^-E_o + t^- && h \in I_o \\ T_i(p_o) &\leq s^+e_i + t^+ && i \in B_o,\ e_i > E_o \qquad (1.14) \\ T_j(p_o) &\leq s^-e_j + t^- && j \in B_o \quad e_j < E_o \end{aligned}$$

Obviously, any $s \in [s^+, s^-]$ generates the point (E_o, C_o).

Since always $C(E,\mathbb{Q}) \leq sE + t_s$ with equality if and only if s generates the point (E,C), we have the characterization

$$C(E,\mathbb{Q}) = \min_{s \geq 0} \{sE + t_s\} \qquad (1.15)$$

According to the normalization of the expense vector e the minimal value of E is $E_{min} = 0$. This point is reached if and only if $e_j = 0$ for all $j \in I$. $C(E_{min},\mathbb{Q})$ is zero if only one e_j is zero, otherwise, the constrained capacity may be positive. In the case $E = E_{min}$ the conditions (1.3) read

$$\begin{aligned} T_j &= t_{min} && j \in I \\ T_j &\leq se_j + t_{min} && j \in B \end{aligned} \qquad (1.16)$$

and this is satisfied for any value $s \geq s_{max}$ where s_{max} is defined as the minimal value such that (1.16) holds.

As an example we give the explicit expression for s_{max} in the binary case. Assume that $e = \binom{e_1}{0}$. For any $\varepsilon > 0$ we have

$$T_1^\varepsilon = s_\varepsilon e_1 + t_\varepsilon$$

$$T_2^\varepsilon = t_\varepsilon$$

or $s_\varepsilon = e_1^{-1}(T_1^\varepsilon - T_2^\varepsilon)$ and with $p_o = \binom{0}{1}$ it follows

$$\lim_{\varepsilon \to 0} s_\varepsilon = s_{max} = e_1^{-1}\left[Q_{11} \log \frac{Q_{11}}{Q_{12}} + Q_{21} \log \frac{Q_{21}}{Q_{22}}\right]$$

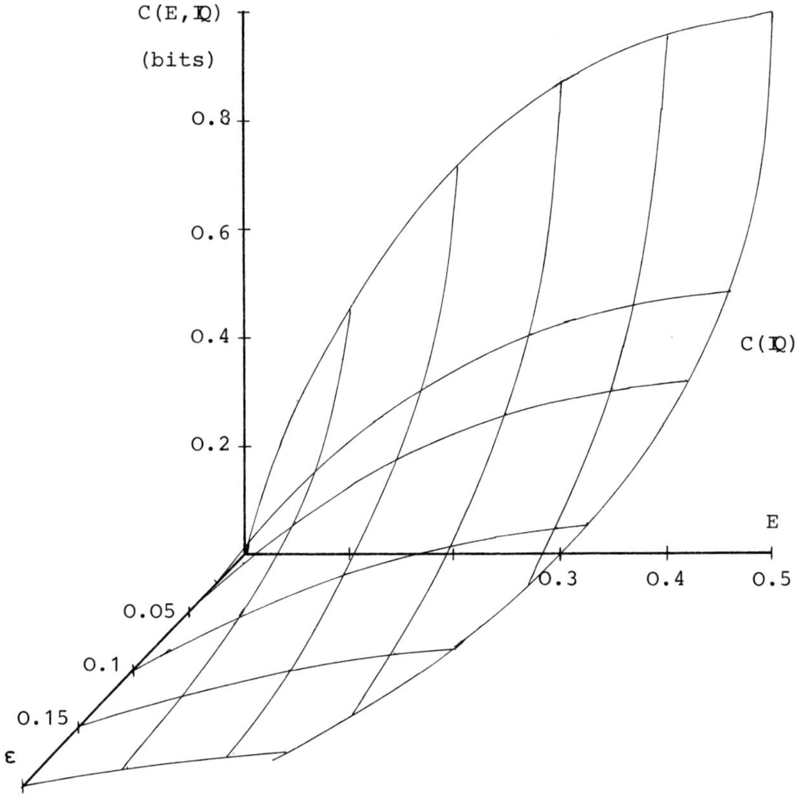

Fig. 1

Capacity-expense surface for $e_1 = 0$, $e_2 = 1$, $Q_{11} = 1-\varepsilon$, $Q_{12} = 2\varepsilon$, $0 \leq \varepsilon \leq 0.2$

Fig. 1 shows an example of a capacity-expense function in the binary case. Note that any matrix $\mathbb{Q}(\varepsilon)$ may be expressed as a convex combination of the matrices $\mathbb{Q}(0)$, $\mathbb{Q}(\varepsilon_0)$, $0 \leq \varepsilon \leq \varepsilon_0$. For $\varepsilon = 0$ we have $Q_{kj} = \delta_{kj}$. It is straightforward to determine the constrained capacity of a lossless channel for arbitrary alphabet sizes. The conditions (1.3) then read $-\log p_j = se_j + t$, $j = 1...L$, or $p_j = \exp(-se_j - t)$. Obviously all p_j are nonnegative. Since $\sum_j p_j = 1$, we can eliminate t and have

$$p_j = \exp(-se_j) \left[\sum_i \exp(-se_i)\right]^{-1}, \quad E = \sum_j p_j e_j$$
$$C(E,\mathbb{1}) = H(p) \tag{1.17}$$

We see from (1.17) that $\lim_{s\to\infty} p_j = 0$, if $e_j > 0$, $\lim_{s\to\infty} p_j = 1$, if $e_j = 0$.

2. The Rate-Distortion Function

In this optimization problem the source is given and one is looking for a channel which operates with minimal transinformation such that a prescribed distortion is not exceeded. The simplest case we are considering here is a discrete, stationary, independently distributed source with probability vector $p = (p_1 \ldots p_L)^t$. We define a distortion matrix $\mathbb{D} = (d_{kj})_{j=1\ldots L, k=1\ldots M}$, $L,M \geq 2$ of real valued elements and have with a transition matrix $\mathbb{Q}$ of same size an average distortion

$$d(p,\mathbb{Q}) = \sum_{j,k} p_j Q_{kj} d_{kj}$$

The convex optimization problem is as follows:

$$R(p,D) = \min_{\mathbb{Q}} T(p,\mathbb{Q}) \tag{2.1a}$$

$$d(p,\mathbb{Q}) \leq D \tag{2.1b}$$

$$\sum_k Q_{kj} = 1 \ , \ j = 1\ldots L \tag{2.1c}$$

$$Q_{kj} \geq 0, \ j = 1\ldots L, \ k = 1\ldots M \tag{2.1d}$$

Since the set $Q_D = \{\mathbb{Q} \mid d(p,\mathbb{Q}) \leq D\}$ is contained in the set $Q_{D'}$, if and only if $D \leq D'$, $R(p,D)$ as function of D is nonincreasing and a solution lies always on the boundary of the convex set Q_D. Therefore we can replace the inequality constraint (2.1b) by $d(p,\mathbb{Q}) = D$. When we ignore temporarily the constraint (2.1d), which is again the main obstacle in a determination of the rate-distortion function, any stationary point $\mathbb{Q}^*$ has to be determined from the Lagrange function

$$L(p,\mathbb{Q}) = T(p,\mathbb{Q}) - sd(p,\mathbb{Q}) - \sum_j t'_j \sum_k Q_{kj}$$

and the requirement

$$\frac{\partial L(p,\mathbb{Q})}{\partial Q_{kj}} = p_j \log \frac{Q_{kj}}{q_k} - sp_j d_{kj} - t'_j = 0 \tag{2.2}$$

for all j,k. With the substitution $t'_j = p_j \log t_j$ this gives

$$Q_{kj} = q_k t_j \exp(sd_{kj}) \tag{2.3}$$

Summing (2.3) over k and observing (2.1c) we obtain an explicite expression for the Lagrange multiplier t_j:

$$t_j = [\sum_k q_k \exp(sd_{kj})]^{-1} \tag{2.4}$$

Multiplying (2.3) by p_j and summing over j yields

$$\sum_j p_j Q_{kj} = q_k = \sum_j p_j q_k t_j \exp(sd_{kj})$$

or

$$c_k := \sum_j p_j t_j \exp(sd_{kj}) = 1 \ , \ q_k > 0 \qquad (2.5)$$

This is a simplification of the problem as (2.4) and (2.5) contain only the smaller set of the output probabilities q_k. From a solution vector q the transition probabilities may be found via (2.3).

If a solution vector has only nonnegative components q_k, the complete problem has been solved. If not, then the solution lies on the boundary and at least one of the components q_k must be zero. With a similar argument as in the previous chapter it can be shown that (2.5) has to be replaced by

$$\begin{aligned} c_k &= 1 \qquad k \in I \\ c_k &\leq 1 \qquad k \in B \end{aligned} \qquad (2.6)$$

Combining (2.3) with the expressions for $T(p,\mathbb{Q})$, $d(p,\mathbb{Q})$, we get the parametric representations

$$\begin{aligned} D &= \sum_{j,k} p_j t_j q_k d_{kj} \exp(sd_{kj}) \\ R(p,D) &= sD + \sum_j p_j \log t_j \end{aligned} \qquad (2.7)$$

Again the usual way to calculate rate-distortion functions is to select a value for the parameter s and then to solve (2.6), (2.7) by analytical or numerical techniques. See Blahut (1972) for a simple and always convergent computer algorithm.

The next theorem states the convexity property of the R(p,D)-function

<u>Theorem 2.1</u>

The rate-distortion function R(p,D) is convex $\cup$ with respect to D and convex $\cap$ with respect to p.

<u>Proof</u>

a) Let $\mathbb{Q}', \mathbb{Q}''$ be two transition matrices such that for a given input vector p $T(p,\mathbb{Q}') = R(p,D')$, $T(p,\mathbb{Q}'') = R(p,D'')$, $D' \neq D''$. The linearity of $d(p,\mathbb{Q})$ implies that $\mathbb{Q}^\beta = \beta'\mathbb{Q}' + \beta''\mathbb{Q}'' \in Q_{\beta'D' + \beta''D''}$ and we have

$$\begin{aligned} R(p,\beta'D'+\beta''D'') &\leq T(p,\mathbb{Q}^\beta) \leq \beta' T(p,\mathbb{Q}') + \beta'' T(p,\mathbb{Q}'') \\ &= \beta' R(p,D') + \beta'' R(p,D'') \end{aligned} \qquad (2.8)$$

b) Let $p', p'', p^\beta, \mathbb{Q}', \mathbb{Q}'', \mathbb{Q}^\beta$ be source and transition probabilities such that p^β is a convex combination of p', p'' and for a given D $R(p',D) =$

$T(p',\mathbb{Q}')$, $R(p'',D) = T(p'',\mathbb{Q}'')$, $R(p^\beta,D) = T(p^\beta,\mathbb{Q}^\beta)$. Then we have

$$\beta' R(p',D) + \beta'' R(p'',D) \leq \beta' T(p',\mathbb{Q}^\beta) + \beta'' T(p'',\mathbb{Q}^\beta)$$
$$\leq T(p^\beta,\mathbb{Q}^\beta) = R(p^\beta,D) \tag{2.9}$$

QED

The next theorem gives a geometrical interpretation of the parameter s. We write $t_s = \sum_j p_j \log t_j$.

Theorem 2.2

Let (D_s,R_s) be a point generated by s for a fixed p. Then $y = sx + t_s$ is tangent line at this point.

Proof

Assume there is another point (D',R') in common with $y = sx + t_s$, that is

$$R' = R_s + s(D' - D_s) \tag{2.10}$$

By definition

$$R' = \min_{\mathbb{Q}\in Q_{D_s}} \{T(p,\mathbb{Q})\} - sD_s + sD' = \min_{\mathbb{Q}}\{T(p,\mathbb{Q}) - sd(p,\mathbb{Q})\} + sD'$$
$$= \min_{\mathbb{Q}}\{T(p,\mathbb{Q}) - s(d(p,\mathbb{Q}) - D')\} \tag{2.11}$$

From (2.10) and (2.11), we see that s generates (D',R') and there is either $D' = D_s$ or s generates a straight line segment between the points (D_s,R_s) and (D',R').

QED

Corollary: As the slope of a nonincreasing function, $s \in [-\infty,0]$.

Any tangent line of a convex ∪ function never overestimates the function and we can write $R(p,D) = \max_{s\leq 0}\{sD + t_s\}$. Furthermore, it is possible to define t_s as a maximum rather than a minimum. To this end let t^*, q^* be solution vectors of (2.6). Then select a vector t such that $c_k = \sum_j p_j t_j \exp(sd_{kj}) \leq 1$. From (2.6) the following chain results

$$0 \geq \sum_{j,k} q_k^* p_j (t_j - t_j^*) \exp(sd_{kj}) = \sum_{j,k} p_j q_k^* t_j \exp(sd_{kj}) \left(\frac{t_j}{t_j^*} - 1\right)$$
$$= \sum_{j,k} p_j Q_{kj}^* \left(\frac{t_j}{t_j^*} - 1\right) \geq \sum_{j,k} Q_{kj}^* p_j \log \frac{t_j}{t_j^*} = \sum_j p_j \log \frac{t_j}{t_j^*}$$

or

$$\sum_j p_j \log t_j \leq \sum_j p_j \log t_j^*$$

and we have

$$R(p,D) = \max_{s\leq 0} \max_{c_k \leq 1} \{sD + \sum_j p_j \log t_j\} \tag{2.12}$$

(2.12) provides us with a convenient lower bound which may be written as

$$R(p,D) \geq H(p) + sD + \sum_j p_j \log p_j t_j \ , \ s \leq 0, \ c_k \leq 1 \qquad (2.13)$$

In the further discussion, we assume without restricting generality the matrix elements d_{kj} to be normalized according to the always possible transformation $d_{kj} = d'_{kj} - \min\{d'_{kj}\}$. Because of the linearity of $d(p,\mathbb{Q})$, this shifts the R(p,D)-curve along the D-axis such that $D_{min} = 0$ with the associated solution $Q_{kj} = 1$ if $d_{kj} = 0$. D ranges between D_{min} and $D_{max} = \min\{D|R(p,D) = 0\}$. Since the transinformation $T(p,\mathbb{Q})$ is zero if and only if input and output are statistically independent, $d(p,\mathbb{Q})$ reduces in this case to $d(p,\mathbb{Q}) = \sum_k q_k \sum_j p_j d_{kj}$. Obviously,

$$D_{max} = \min_k \{ \sum_j p_j d_{kj} \} \qquad (2.14)$$

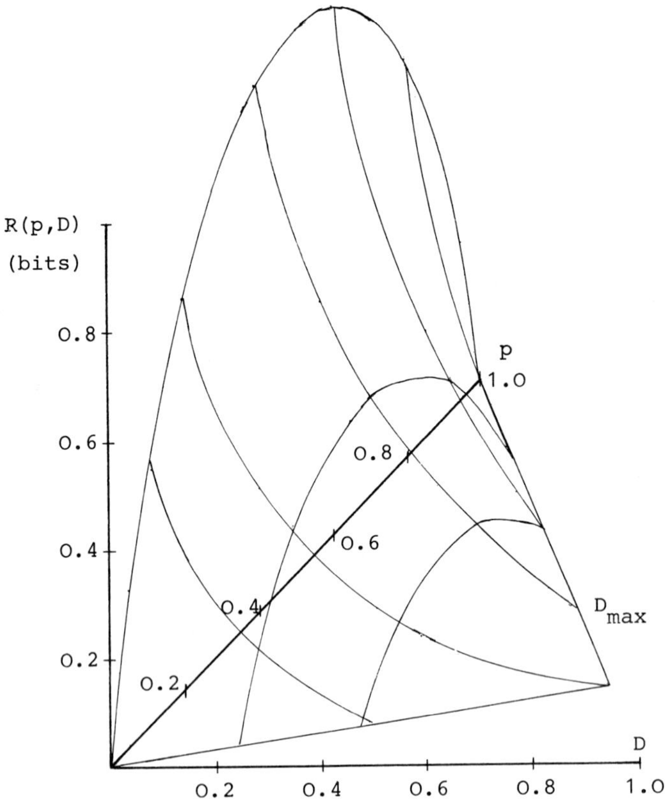

Fig. 2

Rate-distortion surface for $D_{11} = 0 = D_{22}$, $D_{12} = 4$, $D_{21} = 1$, $0 \leq p_2 \leq 1$.

and the set $\{(p,D) \mid R(p,D) \geq 0\}$ is a convex polyhedron.

Fig. 2 shows an example of a rate-distortion surface in the binary case with distortion matrix $\mathbb{D} = \begin{pmatrix} 0 & 4 \\ 1 & 0 \end{pmatrix}$.

As an example, we give the explicite R(p,D)-function for $\mathbb{D} = \begin{pmatrix} 0 & 1 \\ 1 & 0 \end{pmatrix}$. A straightforward calculation via (2.4), (2.5) and (2.7) reveals that

$$R(p,D) = \begin{cases} H(p) - H(D,1-D), & D \leq D_{max} \\ 0 \text{ otherwise} \end{cases} \tag{2.15}$$

Finally, we remark that any rate-distortion function reaches the point R(p,0) with slope $s = -\infty$. As noted above, a solution matrix for $D = 0$ has the property $Q_{kj} = 1$ if $d_{kj} = 0$ and $Q_{kj'} = 0$ for all $j \neq j'$. From (2.3) it is evident that this is impossible for any finite value of s. In contrast to the constrained capacity function, the slope of any rate-distortion curve is continous within the range $0 < D < D_{max}$.

3. Some Examples and Implications

Analytical calculations both of the capacity-expense and the rate distortion function are possible only for simply structured vectors and matrices involved in the optimization problems (1.1) resp. (2.1). In what follows we shall discuss simple examples where a complete analytical determination is not possible but some general properties of the solutions can be found.

I. First let us discuss a generalization of (1.17). Let $\mathbb{Q}$ be a $L \times L$-matrix of the form $Q_{kj} = (1-\delta_{kj})\varepsilon + \delta_{kj}(1-\varepsilon(L-1))$ and e an expense vector of dimension L (assume $\varepsilon < (L-1)^{-1}$). Then we have from (1.3)

$$-\varepsilon \cdot \sum_{k \neq j} \log q_k - (1-\varepsilon(L-1)) \log q_j = h(\varepsilon) + se_j + t, \quad j \in I \tag{3.1}$$

where $h(\varepsilon) = H(1-\varepsilon(L-1),\varepsilon,\ldots\varepsilon)$. The simple form of $\mathbb{Q}$ admits an explicit relation between p_j and q_j: $q_j = p_j(1-\varepsilon L)+\varepsilon$. After adding and subtracting $\varepsilon \log q_j$ on the left side of (3.1), we can write

$$p_j = (1-\varepsilon L)^{-1}\left[\exp\{-(1-\varepsilon L)^{-1}(se_j + t + h(\varepsilon) + \varepsilon\sum_k \log q_k\} - \varepsilon\right], \quad j \in I$$

$$(2\varepsilon L - 1 - \varepsilon) \log \varepsilon + (1-\varepsilon) \log (1-\varepsilon) \leq \varepsilon \sum_k \log q_k + se_j + t, \quad j \in B \tag{3.2}$$

Without restricting generality, we assume the components of e to be ordered in the form $0 = e_1 \leq e_2 \leq \ldots \leq e_L$. Then an immediate consequence of (3.2) is $p_1 \geq p_2 \geq \ldots \geq p_L$. The 'cheaper' inputs are used more frequently which is a reasonable result. If $L > 2$, then depending on e and s the p_j may vanish in the order $p_L, p_{L-1}, \ldots$.

II. A simple distortion matrix is the so called error distortion measure $d_{kj} = 1 - \delta_{kj}$, $j,k = 1...L$. From (2.4) and (2.6) we have

$$t_j^{-1} = \begin{cases} q_j + (1-q_j)\exp s & j \in I \\ \exp s & j \in B \end{cases} \tag{3.3}$$

$$c_k = \sum_j p_j t_j \exp s + (1 - \exp s) p_k t_k = 1 \qquad k \in I$$

Obviously, $p_k t_k$ must take the same value α for all $k \in I$ and this gives $\alpha(q_k + (1-q_k)\exp s) = p_k$ or

$$q_k = \frac{p_k - \alpha\exp s}{\alpha(1 - \exp s)} \qquad k \in I \tag{3.4}$$

Actually, it can be shown via (2.8) and (3.3) that

$$q_k = (p_k - \alpha\exp s)\left[\sum_{l \in I}(p_l - \alpha\exp s)\right]^{-1} \qquad k \in I \tag{3.5}$$

α can be calculated from (3.4). The implicit relation (3.5) shows that $p_1 \geq p_2 \geq \cdots \geq p_L$ implies $q_1 \geq q_2 \geq \cdots \geq q_L$ and the output letters are equiprobable if and only if the source letters are equiprobable. We have $D_{max} = 1 - p_1$, that is $q_1 = 1$ at $D = D_{max}$ and the letters with highest index k vanish first when s increases.

III. Let $\mathbb{D}$ be a distortion matrix with $d_{kj} = j(1 - \delta_{kj})$, $j,k = 1...L$. Then

$$t_j^{-1} = q_j + (1 - q_j)\exp(sj) \tag{3.6}$$

$$c_k = \sum_j p_j t_j \exp(sj) + (1 - \exp(sk)) p_k t_k = 1 \qquad k \in I$$

and by the same argument as above it follows

$$q_k = p_k/\alpha - \alpha\exp(sk)/(1 - \exp(sk)) \qquad k \in I \tag{3.7}$$

If the source letters are equiprobable, that is $p_k = 1/L$, then $q_k > 0$ if $s < -k^{-1}\log(1 + 1/\alpha L)$. This implies that the output letters vanish in the order $q_1, q_2, \ldots$ when s increases. This is in accordance with $D_{max} = \min_k\{\frac{1}{L}\sum_j j(1 - \delta_{kj})\} = \frac{1}{L}\sum_j j(1 - \delta_{Lj})$, that is, $q_L = 1$ at $D = D_{max}$.

IV. An interesting case is the distortion matrix $d_{kj} = k(1 - \delta_{kj})$, $j,k = 1...L$. We have

$$t_j^{-1} = q_j(1 - \exp(sk)) + \sum_l q_l \exp(sl) \tag{3.8}$$

$$c_k = \exp(sk)\sum_j p_j t_j + (1 - \exp(sk)) p_k t_k = 1 \qquad k \in I$$

and this gives with the k-independent constants $\alpha = \sum_j p_j t_j$, $\beta = \sum_l q_l \exp(sl)$

$$q_k = p_k(1 - \alpha \exp(sk))^{-1} - \beta(1 - \exp(sk))^{-1} \qquad k \in I \qquad (3.9)$$

In the simplest case where the source letters are equiprobable, it is easy to see that the output probabilities have the property $q_1 \geq q_2 \geq \ldots \geq q_L$. This is a consequence of $\alpha = \frac{1}{L}\sum_j [q_j + \sum_{l \neq j} q_l \exp(sl)]^{-1} \geq 1$. The condition $q_k > 0$ leads to $\exp(sk) > (L\beta - 1)/(L\alpha\beta - 1)$ which implies that the output letters with highest index k vanish first. This is intuitively reasonable and in accordance with $D_{max} = \min_k \{\frac{1}{L}\sum_j k(1 - \delta_{kj})\} = \frac{1}{L}\sum_j (1 - \delta_{1j})$, that is, $q_1 = 1$ at $D = D_{max}$.

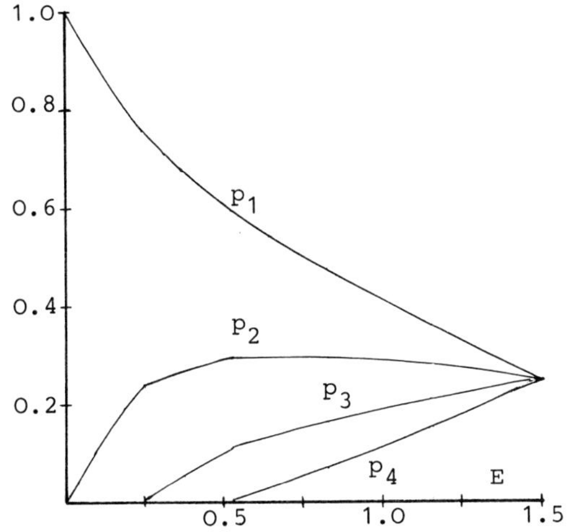

Fig. 3

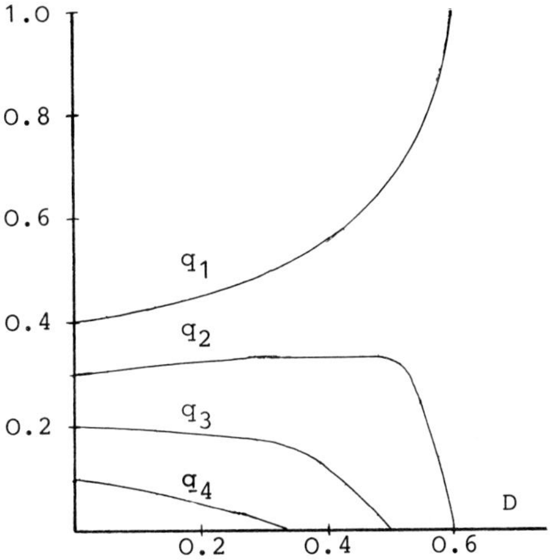

Fig. 4

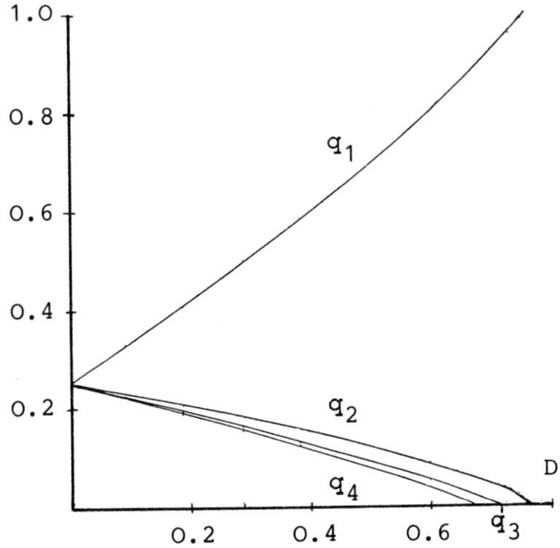

Fig. 5

Fig. 3: Optimal input probabilities of a constrained channel with transition matrix $Q_{kj} = (1-\delta_{kj})\varepsilon + \delta_{kj}(1-\varepsilon(L-1))$, $\varepsilon = 0.1$, $j,k=1\ldots4=L$.

Fig. 4: Optimal output probabilities of a rate-distortion function with distortion matrix $d_{kj}=(1-\delta_{kj})$, $j,k=1\ldots4$, $p^t = (0.4, 0.3, 0.2, 0.1)$.

Fig. 5: Optimal output probabilities with distortion matrix $d_{kj} = k(1-\delta_{kj})$, $j,k=1\ldots4$, p equidistributed.

Note the discontinuity of the first derivative at the points where a component vanishes.

By replacing k by f(k) in all formulas of the last example, we obtain the implicit output probabilities for the distortion matrix $d_{kj} = f(k)(1 - \delta_{kj})$. Without restricting generality, we may assume $f(1) = 1$ and f nondecreasing. Fig. 6 shows an example with $f(k) = 2^{k-1}$ and equiprobable inputs. Note the "merging" of the components with high index f(k) which is an obvious consequence of the modified equation (3.9). We

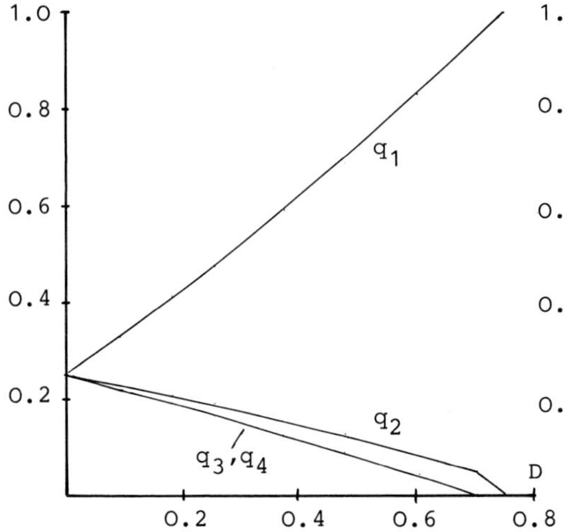

Fig. 6

Optimal output probabilities with distortion matrix $d_{kj} = 2^{k-1}(1-\delta_{kj})$, j,k= 1...4, p equidistributed.

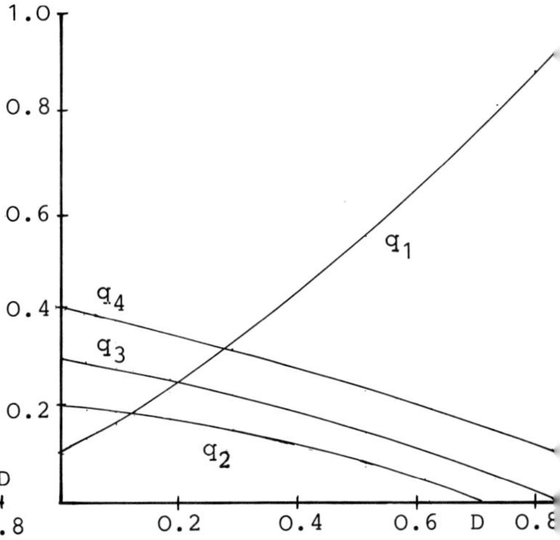

Fig. 7

Optimal output probabilities with distortion matrix $d_{kj} = k(1-\delta_{kj})$, j,k = 1...4, p^t = (0.1, 0.2, 0.3, 0.4).

have for an arbitrary input vector p $D_{max} = \min_k \{ f(k) \sum_j p(j)(1 - \delta_{kj})\} = \min_k\{f(k)(1 - p_k)\} := f(k^*)(1 - p_{k^*})$, that is, $q_{k^*} = 1$ at the point $D = D_{max}$ depending both on f and p. See Fig. 7.

In general, the behavior of the components q_k as functions of D is very complex and only to find out by numerical techniques. It is possible that a component vanishes and comes back into the interior set several times. See Bergér (1972) for some analytically treatable examples.

The matrix $\mathbb{D}$ with components $d_{kj} = f(k)(1 - \delta_{kj})$ as defined above expresses that an incorrect mapping from an input to an higher indexed output gives rise to more distortion. The fact that the higher indexed output components vanish first is perhaps easier to see when we adopt another point of view: Interchanging the dependent and the independent variable of the rate-distortion function, we get the equivalent distortion-rate function by looking for an optimal transition matrix $\mathbb{Q}$ such that the average distortion D is minimized for a given transinformation. If all inputs have the same weight, it is natural to cancel the outputs with highest distortion first. The order of vanishing is unaffected if the input components are $p_1 \geq p_2 \geq \dots \geq p_L$. This follows from the general condition $\exp(sf(k)) \geq (\beta - p_k)(\alpha\beta - p_k)$ if and only if $k \in I$. In general, different input probability vectors p cause different interior sets I_p for intermediate values $D > 0$.

By defining, for example, a utility matrix $\mathbb{U}$ rather than a distortion matrix $\mathbb{D}$, we obtain a rate-utility function $R(p,U)$ simply by replacing the constraint (2.1b) by $u(p,\mathbb{Q}) = \sum_{j,k} p_j Q_{kj} u_{kj} \geq U$. Since after normalization a high utility corresponds to a low distortion and vice versa, a rate-utility function is simply a mirror image of a rate-distortion function, that is monotonic increasing with U, $R(p,U=0) = 0$, $R(p,U_{max}) = \max_U \{R(p,U)\}$.

An example where a utility constraint is more appropriate is an information theoretic interpretation of feature detectors in sensory pathways. Feature detectors select the most significant, most utile data and reduce the necessary channel capacity between sensory input and brain. This corresponds to a point on a rate-utility surface where the interior set I is smaller than the set of the inputs. If flexibility in a varying environment is required, the channel outputs are a union of several interior sets $I_1, I_2, \dots I_n$ for an interior set I is optimal only for a certain input distribution and a neighbourhood. Thus the data compression is smaller resp. the necessary channel capacity in the sensory pathway higher.

Pfaffelhuber (1974) discussed minimum cost noiseless coding in the context of optimal spikes firing rates. The optimal source resp. channel input distribution is given in (1.17). In contrast to this special case where all probabilties are nonzero, the optimal input distribution of a noisy channel may have zero components (see Fig. 3) for lower values of the expense E. Since the channel can be reduced by that unused input letters, the transmission complexity may be diminished considerably.

Both the capacity-expense and the rate-distortion functions exhibit an interesting behavior near its maximal values. The point $C(E_{max}, \mathbb{Q})$ of any capacity expense function is always reached with Lagrange parameter $s = 0$, that is, with slope zero. This implies that the last part of the channel capacity requires a disproportionate amount of expense E. This is a phenomenon which may be observed in various contexts.

The point $R(p,0)$ of any rate-distortion function is always reached with Lagrange parameter $s = -\infty$. This implies that even a small amount of tolerable distortion resp. a small amount of tolerable loss of maximal possible utility may cause a dramatic reduction in necessary channel capacity which has great practical significance as transmission costs vary generally directly with capacity. The larger the input alphabeth is, the more marked is this effect of a initially fast decreasing rate-distortion function. When the input alphabeth is countably infinite for instance, in general $R(p,D)$ diverges for $D \to 0$, but is finite and gives rise to a finite interior set I for any $D > 0$.

As an example, we roughly lower bound $R(p,D)$ for the distortion matrix $d_{kj} = f(k)(1 - \delta_{kj})$ using (2.13). With $D_{max} = \min_k \{f(k)(1 - p_k)\}$ we choose $s = \log D/D_{max}(1 - D)$ which implies $-\infty \leq s \leq 0$ if $0 < D < D_{max}$. To have a rough estimate for $p_j t_j$, we take $p_j t_j$ = const. and get from (3.8)

$$\log p_j t_j = - \log \left(1 + \left(\frac{D}{D_{max}(1-D)}\right)^{f(j)}\right) \geq - \left(\frac{D}{D_{max}(1-D)}\right)^{f(j)}$$

which gives

$$R(p,D) \geq H(p) + D\log \frac{D}{D_{max}(1-D)} - \sum_j p_j \left(\frac{D}{D_{max}(1-D)}\right)^{f(j)} \tag{3.10}$$

Finally, we remark that it is straightforward to introduce both in (1.1) and (2.1) multiple constraints as a set of expense vectors e^i resp. a set of distortion matrices $\mathbb{D}^i$. A simple calculation reveals the parametric representations

$$C(E^1,E^2,\ldots E^n,\mathbb{Q}) = \sum_i^n s^i E^i + t$$

resp.

$$D^i = \sum_{j,k} p_j t_j q_k d^i_{kj} \exp\left(\sum_i^n s^i d^i_{kj}\right) \tag{3.11}$$

$$R(p,D^1,D^2,\ldots D^n) = \sum_i^n s^i D^i + \sum_j p_j \log t_j$$

Properties of multiple constrained transinformation functions are discussed in a subsequent paper.

Acknowledgments

The autor is grateful to Prof. V. Braitenberg and to Dr. G. Palm for many stimulating discussions and helpful comments.

References

Barlow, H.B.: Redundancy and Perception, Springer Lecture Notes in Biomathematics, 4, pp. 458, 1974

Berger, T.: Rate Distortion Theory, Prentice Hall, New Jersey, 1971

Blahut, R.E.: Computation of Channel Capacity and Rate-Distortion Functions, IEEE Trans. Inform. Theory, IT-18, pp. 460 - 473, 1972

Gallager, R.G.: Information Theory and Reliable Communication, Wiley, New York, 1968

Heim, R.: Zufall und Information, doctoral thesis, University of Tübingen, Tübingen, 1976

Heim, R.: On the Algorithmic Foundation of Information Theory, submitted to IEEE Trans. Inform. Theory, 1976

Pearl, J.: Theoretical Bounds on the Complexity of Inexact Computations, IEEE Trans. Inform. Theory, IT-22, pp. 580 - 586, 1976

Pfaffelhuber, E.: Sensory Coding and the Economy of Nerve Impulses, Springer Lecture Notes in Biomathematics, 4, pp. 467 - 483, 1974

Self-Diagnosis of Interactive Systems

M. Dal Cin
Institute for Information Sciences
University of Tübingen

Dedicated to Ernst Pfaffelhuber

1. Introduction

J. von Neumann |1| provided the first rigorous study of fault tolerance in complex systems. He analyzed the role of fault masking redundancy in (formal) neural networks. Soon he realized that static fault tolerance alone cannot explain the high reliability of complex biological systems such as the brain. He was led to assume that "the ability of a natural organism to survive in spite of a high incidence of error (...) probably requires a very high flexibility and ability (...) to watch itself and reorganize itself. And this probability requires a very considerable autonomy of parts." (von Neumann |2| p. 73). Fault diagnosis, that is fault detection and location, is necessary before any action against faults can be taken. Following von Neumann, we view a fault tolerant system as a collection of interacting subsystems (units) which are capable of testing (and repairing) other subsystems and of being tested.

Because of the expanding application of computers into areas requiring high availability self-testing is becoming an important issue of system design |3|. Multiple processor systems with spare resources available have been recently proposed where each processor may enter into the diagnosis and reconfiguration of other processors. A popular multiprocessor architecture is the ring structure |4|, c.f. Fig. 1. Processor J periodically tests processor J+1 to determine its health. The affected processor simply removes itself from the loop and a standby may be switched in while the faulty processor is being repaired. However, the self-testing ability is not peculiar to computing systems, it can be observed in biological and socioeconomic systems as well.

Studying the performance of self-testing systems two problems arise.
(1) In a self-testing system conflicts may arise between the testing mode of the units and their regular service mode. High service load can cause insufficient response time for testing. The system, designed for

self-testing, would not test itself. On the other hand, if too many tests are performed, severe contention problems will occur with a resulting degradation in the regular performance of the system. Therefore, one must try to strike a reasonable balance of service. To this end, we have to respond to questions like: when will the first test of the whole system be completed, what portion of system time (in the stationary state) will be taken up by testing and how will this portion vary as a function of the system's parameters?

(2) The test outcomes reported by a good unit are reliable but those reported by a faulty unit are meaningless. Hence, in most self-testing systems the faulty units cannot be unambiguously identified. Of course, this ambiguity has to be reduced as much as possible. We are led, therefore, to consider models with a limited network topology with respect to test links.

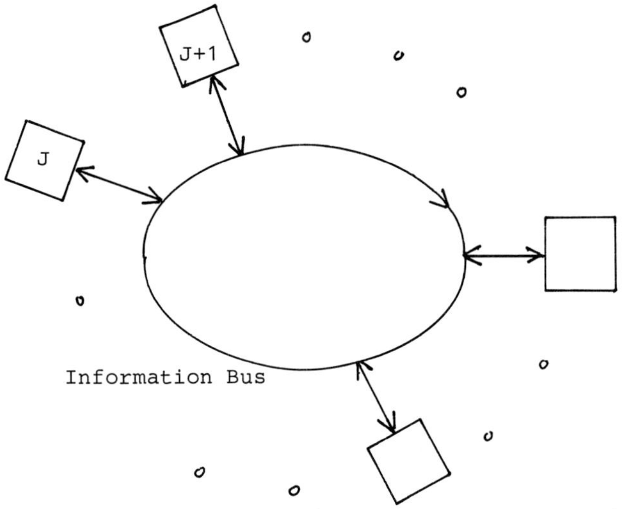

Fig. 1 Multi-processor ring

Our investigation is based on the wellknown graph theoretical model of self-testing systems proposed by Preparata, Metze, and Chien |5|. In Sec. 2 we recall the definition of t-diagnosable and sequentially t-diagnosable self-testing systems and discuss briefly the design of optimal test graphs. In Sec. 3 we introduce a new diagnosability measure, viz. probabilistic diagnosability. In order to study the performance of self-testing systems we then introduce two different modes of testing, viz. testing by "interviews" (test mode A) and testing by "questionairs" (test mode B); Sec. 4. The measures of interest are system throughput and diagnosability. Two illustrative examples are discussed in Sec. 5.

2. Self-Diagnosis

Example: Each participant in a round table discussion states whether or not his right neighbor seems to be competent. There are five discussants and the sequence of judgements (starting with the discussion leader Dr. H) is 10101. Who is competent, who is not?

Consider a system Σ consisting of n units σ_i (i=1,...,n) each of which is capable of applying tests to other units and/or being tested by them. This testing arrangement is represented by a test graph $G_\Sigma = (V,E)$ where the set of vertices, V, represents the units σ_i of Σ and $E \subset V \times V$ represents the set of test links |5|. That is, unit σ_i tests unit σ_k iff $(\sigma_i,\sigma_k) \in E$. The test outcome, t_{ik}, is 1 (good) or 0 (faulty); t_{ik} is the judgement of unit σ_i on unit σ_k. The units may be programs, microprocessors, computers (e.g. in a multi microcomputer system) or people. Unit σ_i is said to be in state $q_i=1$ (up) if it is functioning; it is said to be in state $q_i=0$ (down) if it is faulty. The tuple $\Gamma = (t_{ik})$ of all test outcomes at a certain test instance is called a syndrom of Σ. The test outcomes reported by a good unit are reliable, but those reported by a faulty unit are meaningless. Therefore, the problem is to decode the syndroms. That is, given a syndrom Γ, how to identify all faulty units. Given a syndrom $\Gamma \in \{0,1\}^{|E|}$, a fault pattern $Q = (q_1,q_2,\ldots,q_n)$, $q_i \in \{0,1\}$, is said to be consistent with Γ if Σ can produce syndrom Γ if unit σ_i is in state q_i, i=1,2,...,n.

For most syndroms and test graphs the faulty units cannot be unambiguously identified unless additional information on the consistent fault patterns is available. Suppose, the maximum number of faulty units is known. Preparata et. al. call a self-testing system t-diagnosable (t-db) iff for each syndrom of Σ all faulty units can be uniquely identified provided there are no more than t faulty units in the system. They call a self-testing system sequentially t-diagnosable (s-t-db) iff at least one faulty unit can be identified. If Σ is sequentially t-diagnosable, a sequence of at most t test- and repair actions identifies and repairs all faulty units.

Theorem |6|: System Σ, consisting of n units, is t-db iff
(a) $t \leq \frac{1}{2}(n-1)$
(b) each unit of Σ is tested by at least t other units and
(c) for every subset W of V with $|W| = n-2t+\ell$ $(0 \leq \ell \leq t-1)$ the inequality

$$r := |pr_2 [E \cap (W \times (V - W))]| \geqslant \ell + 1 \qquad (1)$$

holds. Relation (a) is also necessary for s-t-diagnosability.

The consistent fault patterns of $\Gamma = (t_{ik})$ can be found by solving the equation:

$$\prod_E [1 - t_{ik}q_i + (2t_{ik} - 1)q_iq_k] = 1 \qquad (2)$$

where $q_i \in \{0,1\}$. That is, for $t_{ik} = 1$ (0) either σ_i is faulty or $q_k = 1$ (0).

Example: Round table discussion.
Fig. 2 shows the test graph G^1_Σ of the round table, Fig. 3 shows an extended test graph G^2_Σ. Test graph G^1_Σ is 1-db and s-1-db, G^2_Σ is 1-db and s-2-db. (For each $W \subset V$ of 3 units: $\ell=0$, $r \geqslant 1$).
Consistent fault patterns are given by

$$G^1_\Sigma : (1-q_H+q_Hq_S)(1-q_Sq_E)(1-q_E+q_Eq_F)(1-q_Fq_G)(1-q_G+q_Gq_H) = 1$$

	q_H	q_S	q_E	q_F	q_G
"Minimal" solutions are:	1	1	0	0	1
	1	1	0	1	0
Other solutions:	0	1	0	1	0
	0	0	1	1	0
	1	1	0	0	0

At least two participants are not competent. They cannot be identified.

$$G^2_\Sigma : (1-q_H+q_Hq_S)(1-q_G+q_Gq_H-q_Fq_H)(1-q_Eq_S)(1-q_E) = 1$$

Hence, $q_E \equiv 0$ and E is not competent.

Now, let Σ have n units. Test graph $G_\Sigma = (V,E)$ is optimal diagnosable if Σ is t-diagnosable, with $t = \lfloor \frac{n-1}{2} \rfloor$ and each unit is tested by exactly t units. In |5| optimal designs for test graphs are given. Fig. 4 shows an optimal diagnosable test graph. However, the implementation of test links is costly and $|E| = \frac{n(n-1)}{2}$ (n odd) links (and tests) are required for optimal design. Therefore, it is reasonable to keep $|E|$ as small as possible, particularly, if the availability of each unit is high. Also, in real systems not every unit is capable of testing every other unit. In these cases one cannot evade the uncertainty inherent in self-diagnosis.

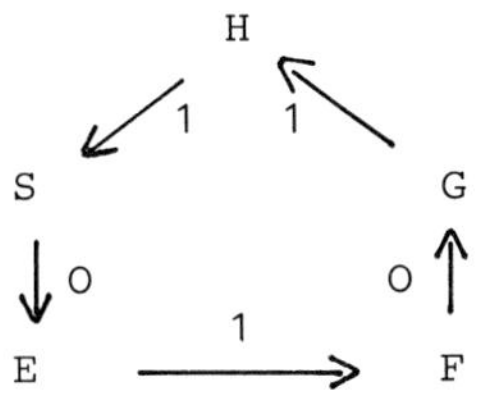

Fig. 2 G^1_Σ

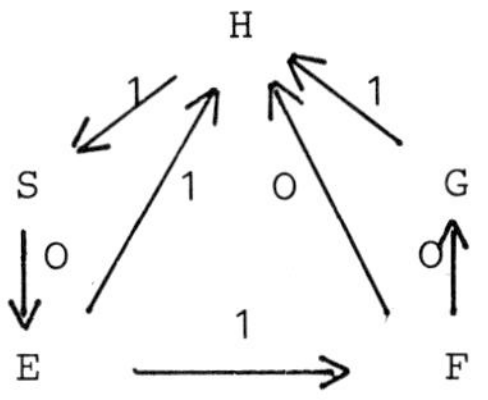

Fig. 3 G^2_Σ

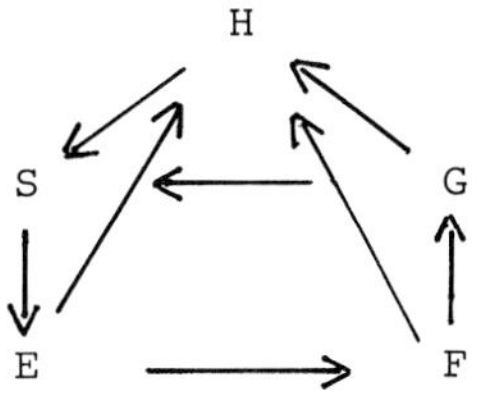

Fig. 4 Optimal test graph

3. Probabilistic Self-Diagnosis

Self-diagnosis can be considered as a problem of pattern recognition. The situation is basically probabilistic. A 0-1 pattern (syndrom) has to be assigned to a pattern class (fault pattern) where complete prior knowledge of the actual pattern class is not available. In order to decode a given syndrom any reasonable decision rule may be applied. We will utilize the following decoding procedure.

Let us assume that the availability $A_i(t) = \Pr\{q_i = 1 \text{ at time } t\}$ is known for each unit σ_i $(0 < A_i(t) < 1)$. Let $|\Gamma|$ be the set of all fault patterns consistent with Γ and assume that the units of Σ are statistically independent with respect to their malfunctioning. The (a posteriori) probability of the occurence of fault pattern $Q = (q_i)$, given that Σ produces syndrom Γ at test time t, is:

$$p_t(Q|\Gamma) = \frac{p(\Gamma|Q)p_t(Q)}{p_t(\Gamma)} = c^{-1}\, 2^{QS} \prod_{i=1}^{n} (1-A_i(t))^{1-q_i}\, A_i(t)^{q_i}$$

if $Q \in |\Gamma|$ and 0 else, where c is such that $\sum_{Q\in|\Gamma|} p_t(Q|\Gamma) = 1$. Here we made the additional assumption that the test outcome of a faulty unit is 0 or 1 with equal probability. Hence, $p(\Gamma|Q) = 2^{(Q-\underline{1})S}$ if $Q\in|\Gamma|$, where S is the column vector $(s_i)^+$ and s_i the out-degree of node σ_i in G_Σ.

Let us further assume that the fault behavior and repair mechanism of each unit σ_i are such that its availability reaches a stationary value A_i (before a catastrophic breakdown occurs) and that the length T_i of time that the unit σ_i is continuously functioning (lifetime) and its repair time W_i are independent random variables. Given these assumptions, renewal theory |7| shows that

$$A_i = \lim_{t\to\infty} A_i(t) = \frac{E[T_i]}{E[W_i]+E[T_i]}$$

where $E[T_i] =: \lambda_i^{-1}$ and $E[W_i] =: \rho_i^{-1}$ is the expected lifetime and repair time of σ_i, respectively. Hence,

$$\lim_{t\to\infty} [\ln p_i(Q|\Gamma)] = \sum_i q_i \ln 2^{s_i} \frac{A_i}{1-A_i} + \bar{c} = \sum_i [\ln 2^{s_i} \frac{\rho_i}{\lambda_i}] q_i + \bar{c}$$

where $Q\varepsilon|\Gamma|$ and $\bar{c}$ is independent of Q. Thus, in order to find the most probable fault pattern, given Γ and stationarity, one has to maximize the function

$$G(Q) = \sum_{i=1}^{n} [\ln 2^{s_i} \frac{\rho_i}{\lambda_i}] q_i \qquad (3)$$

under the constraints (2). The problem of maximizing a function $\{0,1\}^n \to \mathbb{R}$ with constraints is known as pseudo-Boolean programming. Several nonenumerative algorithms are available |8|. Note, that we are using a zero-one loss function. If repair costs matter, the Hamming distance of the actual and the estimated fault pattern is a more realistic loss function.

We are particularly interested in test graphs which admit a unique decoding such that the diagnosis can be done automatically (self-diagnosis).

Definition: System Σ with test graph G_Σ is (stationary) probabilistic diagnosable (p-db) if for all syndroms of Σ there is one and only one consistent fault pattern which maximizes (3). System Σ is sequentially probabilistic diagnosable (sp-db) if for all syndroms and all fault patterns $Q^k = (q_i^k) \neq \underline{1}$ which maximize (3) there is at least one unit σ_i such that $\sum_k q_i^k = 0$.

Thus, we consider the diagnosability of a self-testing system during its usefull lifetime - that is, after early (systematic) faults are eliminated and before complete breakdown. Other measures of diagnosability hav been given in |9|. Fig. 5 shows a decoding unit (switching network).

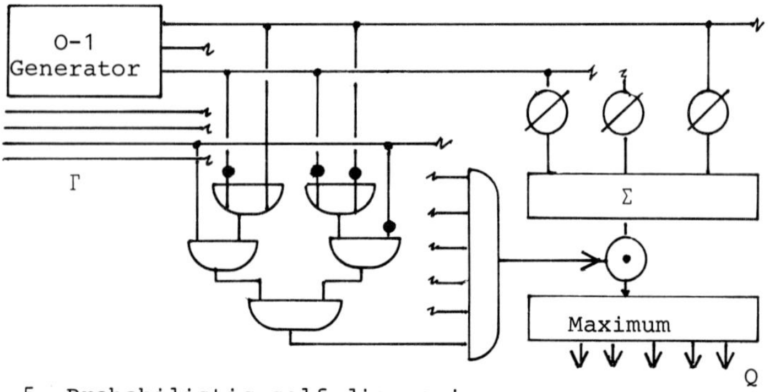

Fig. 5 Probabilistic self-diagnosis

Notation: G^{Γ} denotes the maximum of G(Q) under the constraints (2), $\Gamma = (t_{ik})$. A subset U of V will be identified with the fault pattern (q_i) where $q_i = 1$ iff $\sigma_i \in U$.

Theorem: Given system Σ with test graph $G_{\Sigma} = (V,E)$ and $A_i > 0{,}5$ (i=1, 2,...,n), let Π be the set of all partitions $\pi = (U,X,Y,Z)$ of V with:

$$E \cap [U \times (X \cup Y)] = \emptyset \qquad (4)$$

(where $U = \emptyset = Z$ is allowed). System Σ is not p-db iff there is a $\pi \in \Pi$ such that

$$G(X) = G(Y) = G^{\Gamma_\pi} - G(U). \qquad (5)$$

Syndrom Γ_π is given in Fig. 6. System Σ is not sp-db if, in addition, $Z = \emptyset$.

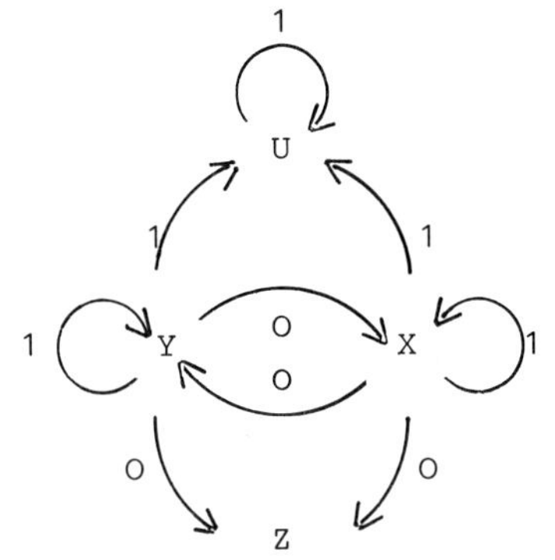

Fig. 6 Syndrom Γ_π. An edge represents a bundle of test links; test links of E not shown may have arbitrary test outcomes.

Proof: From (5) and Fig. 6 follows that $F_1 = U \cup X \in |\Gamma_\pi|$ and $F_2 = U \cup Y \in |\Gamma_\pi|$, $F_1 \neq F_2$ and $G(F_1) = G(F_2) = G^{\Gamma_\pi}$. Hence, Σ is not p-db. If $Z = \emptyset$ then $\bar{F}_1 \cap \bar{F}_2 = \emptyset$ and Σ is not sp-db.

Assume that Σ is not p-db. There are at least one syndrom Γ and two consistent fault patterns F_j, j=1,2, such that $G(F_j) = G^{\Gamma}$. Obviously, $\bar{F}_1 - \bar{F}_2 \neq \emptyset$ and $\bar{F}_2 - \bar{F}_1 \neq \emptyset$, since $A_i > 0{,}5$, i=1,...,n. Define $X=\bar{F}_1-\bar{F}_2$, $Y = \bar{F}_2 - \bar{F}_1$, $Z = \bar{F}_1 \cap \bar{F}_2$ and $U = F_1 \cap F_2$. Thus, $F_2 = X \cup U$ and $F_1 = Y \cup U$. The assumption $(\sigma_i,\sigma_k) \in E \cap [U \times (X \cup Y)]$ implies $t_{ik} = 0$ and $t_{ik} = 1$. Hence, the equations (4) and (5) follow.

Corollary: If Σ is not sp-db then it is not p-db. Hence, (4) and (5) are necessary for non sp-diagnosability. The assumption $A_i > 0{,}5$ can be dropped since there is no sequence $\bar{F}_{j_1} \subset \ldots \subset \bar{F}_{j_2} \subset \ldots \subset \bar{F}_{j_k}$ with respect to all solutions of (3), otherwise Σ is sp-db. If all units have the same (stationary) availability the theorem gives the maximal number of units which may be down, and Σ is still p-db.

Clearly, in p-db systems not every syndrom will be correctly decoded. Therefore, we define the diagnosability D of Σ (for any decoding procedure) as:

(Probability that a test outcome is correctly decoded) x (mean number of tests completion per unit and unit time)

4. Self-Testing Systems as Queueing Networks

In order to predict the (testing) performance of different units of a self-testing system Σ it is natural to represent Σ by a queueing network N_Σ with two classes of jobs: tests (class T) and regular jobs (class R).

A job of class C for unit σ_i characterized by a service time with mean μ_{iC}^{-1} after which the job enters the σ_j-queue with probability $p_{ij}(C)$. (It is assumed that the matrices $(p_{ij}(C))$ are irreducible.) Under heavy traffic conditions we may approximate the system by a closed queueing network with a fixed number, n_R, of regular jobs |10|. That is, we assume that, when a regular job terminates, it enters the "New Job" path (with probability $0 < 1-p < 1$) and the next job is generated. There is a fixed number, n_T, of test jobs in the system. Test jobs are routed according to the test graph G_Σ of Σ.

Two different modes of self-testing will be considered.
Test mode A (interviews): Suppose, unit σ_i is ready for testing unit σ_k. In test mode A unit σ_i waits until unit σ_k is idle. Then both units enter the testing activity (interview). Hence, the service time parameters depend on the overall system's state, e.g., on the number of jobs in a subnetwork of N_Σ.
Test mode B (questionairs): In this mode the completion of a test is represented by a departure of a new test request (a questionair) from the tested unit and the return of the completed test (filled out questionair) to the testing unit. Questionairs wait in queues for service together with regular jobs. Hence, the service time parameters for a test depend only on the state of the unit serving it.

Clearly, the user of a self-testing system Σ is mainly interested in the completion of regular (user) jobs. Thus, the throughput $R_{n_R n_T}$ of Σ is defined as the number of regular jobs completed per unit time. There are n_R regular and n_T test jobs in the system. System throughput and unit repair time, in general, decrease with increase in self-testing (increase in D and n_T). Therefore, we propose as figure of merit of a self-testing system Σ with n_R regular and n_T test jobs the geometric mean:

$$L_{AD}^{n_T} := \left[(A\, R_{n_R n_T} + D_{n_R n_T})\ A\ R_{n_R n_T}\right]^{\frac{1}{2}} \qquad (6)$$

taking into account the (stationary) system availability A. Observe that (a) $L_{AD}^{n_T} = 0$ for $R_{n_R n_T} = 0$ or $A = 0$, and (b) $L_{10}^{0} = R_{n_R 0} \geqslant L_{10}^{n_T}$. Relation (b) expresses the fact, that testing consumes system time.

Now, $D_{n_R n_T}$ is given by

$$D_{n_R n_T} = \frac{B}{n} \sum_{i=1}^{n} T^i_{n_R n_T} \qquad (7)$$

where $T^i_{n_R n_T}$ is the σ_i-throughput of test jobs and B the probability for successful decoding. When we employ the decoding procedure of Sec. 3 then

$$B = \sum_{\Gamma} p(Q^{\Gamma}|\Gamma)\ p(\Gamma) = \sum_{\Gamma} p(\Gamma|Q^{\Gamma})\ p(Q^{\Gamma}) \qquad (8)$$

$$= 2^{-1S} \sum_{\Gamma} \exp\left[\sum_{i=1}^{3} \ln \frac{\lambda_i}{\lambda_i+\rho_i} + G(Q^{\Gamma})\right]$$

where Q^{Γ} is a solution of (3) maximizing B.

To illustrate the model we will apply it to examples. Of course, these special examples do not reveal the entire dynamics of self-testing systems but they do give one an intuitive feeling about their general properties. For further results on the dynamic behavior of self-testing systems the reader is referred to a forthcoming paper |11|.

5. Examples

Let us consider the ring structure (Fig. 1), where unit σ_1, say, is the input-output unit. Its test graph G_{Σ} is given in Fig. 2 (n=5); G_{Σ} is minimal with respect to $|E|$ and sequentially t-diagnosable where $t < 2(\sqrt{n-1}-1)$ if t is even and $t < 2(\sqrt{n-3/4}-1)$ if t is odd. For n=3, G_{Σ} is optimal. Queueing network N_{Σ} represents a self-testing system with fully distributed testing facilities |12|.

5.1 Test mode B (interviews)

We analyze the performance of system Σ in test mode B by means of a

Markov process with velocity dependent state-transition rates |13|.

At any time a unit of system Σ is either active or it is passive. E.g., if σ_i tests an other unit it is in an active state, S_T^i; if it is being tested it is in a passive state. Every state S of the overall system is characterized by real numbers C_S^i. If the system is in state S and unit σ_i is active then C_S^i is the processing speed of σ_i. For example, $C_S^i = 0$ and $S_j^i = S_T^i$ means that testing by σ_i is interrupted since a job with higher priority is in line. The state-transition rates $p(S,S_j^i,\bar{S})$, where S_j^i is an active state of σ_i, specify the flow of jobs within the system. Specifically, $p(S,S_j^i,\bar{S})$ is the probability of a transition from state S into state $\bar{S}$ given that unit σ_i completes its job. Assumption: In every system state at least one unit is active with positive speed.

With the adoption of exponentially distributed service time parameters this model can be represented by a Markov chain. More important, by techniques given in |13| the model is amenable to a sensitivity analysis. That is, one can search for conditions for the processing speeds such that the equilibrium state distribution is the same as in the Markov case (for given expected service times) even so all or some of the service rates are not exponentially distributed. The equilibrium state distribution is then optimized by adjusting the parameters C_j^i under the given constraints. Optimization with respect to processing speeds may be utilized to allocate the system resources (e.g., fast and slow storage media) in such a way that the expected response time for jobs of different classes are met reflecting the importance placed on each class.

Active unit states are: S_T^i <-> unit σ_i is testing, S_E^i (S_F^i) <-> unit σ_i is serving one (two) regular job(s); passive states are: S_t^i <-> unit σ_i is being tested, and S_W^i <-> unit σ_i is idle.

The parameters are given in Tab. 1 and Fig. 7 shows the state-transition graph for n=3. For example, observe that $p(2,2_E^2,3) = p_{23}(R)$ and $p(2,2_E^1,6) = p_{31}(T)$. Tab. 2 gives the equilibrium distribution P_i, i= 1,2,...,9. From this we obtain:

$$R_{n_R n_T} = (1-p)\ \mu\ (P_1+\alpha P_2+\alpha P_3) = \frac{(1-p)\mu\tau\beta}{p\gamma}\ (\beta+2p\alpha) \tag{9}$$

and

$$T_{n_R n_T} = \frac{1}{n} \sum_{i=1}^{3} T^i_{n_R n_T} = \frac{\tau}{3}\ (\delta P_5+\delta P_6+P_7) = \frac{\mu\tau\beta}{3\gamma}\ (\beta+2p\alpha) \tag{10}$$

Since Σ is a (non redundant) series system, its availability is

$A = \prod_{i=1}^{3} \frac{\rho_i}{\lambda_i+\rho_i}$. Tab. 3 shows two numerical examples and the system throughput as no testing takes place. In example 2 - 4 the testing and/or processing speeds of σ_1 and σ_3 are doubled. We assumed that A is independent of $T_{n_R n_T}$.

class R	class T	c_S^i
$p_{23}(R)=p_{31}(R)=1$	$p_{ii+1(3)}(T)=1$	$c_1^1=c_7^2=1$
$p_{12}(R)=1-p_{11}(R)=p$	$i=1,2,3$	$c_2^1=c_3^1=\alpha$
$p_{ij}(R)=0$ else	$p_{ij}(T)=0$ else	$c_2^2=c_3^3=c_4^2=c_4^3=\beta$
$\mu_i(R)=\mu$	$\mu_i(T)=\tau$	$c_5^3=c_6^1=\delta$
$n_R=2$	$n_T=1$	$c_8^2=c_9^3=\varepsilon$
		$c_{i+4}^i=0 \quad i=1,2,3$

Tab. 1 Parameters

S	s_j^1	s_j^2	s_j^3	P_i / γ
1	F	w	w	$p^{-1}\tau\beta^2$
2	E	E	w	$\tau\beta$
3	E	w	E	$\tau\beta$
4	w	E	E	$p\alpha\tau$
5	E	t	T	$\mu\beta^2\delta^{-1}$
6	T	E	t	$p\alpha\mu\beta\delta^{-1}$
7	t	T	E	$p\alpha\mu\beta$
8	w	E	w	$p\alpha\tau\beta\varepsilon^{-1}$
9	w	w	E	$p\alpha\tau\beta\varepsilon^{-1}$

$$\gamma^{-1} = \Sigma \frac{P_i}{\gamma}$$

Tab. 2 State asignment and equilibrium distribution

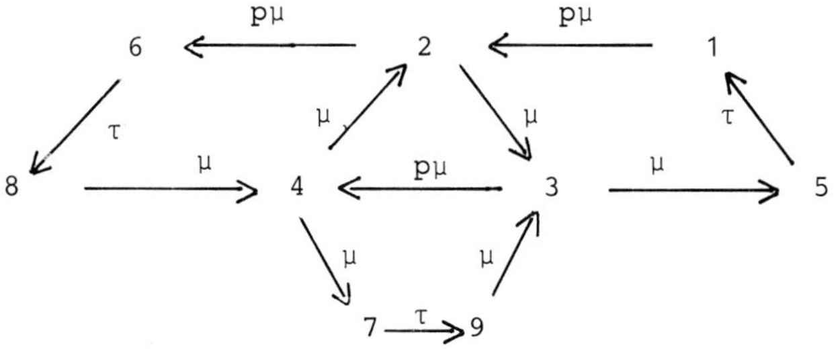

Fig. 7 State-transitions

5.2 Test mode B (questionairs)

The queueing network N_Σ is a cyclic server model |10,12| with two job classes. Let each unit of Σ be complete |14| then the equilibrium state probabilities of N_Σ will have product form. Specifically, let k_{iC} denote the number of jobs of class C at unit σ_i; let $\underline{k}_i = (k_{iR}, k_{iT})$ and $\underline{k} = (\underline{k}_1,\ldots,\underline{k}_n)$. The equilibrium probabilities $(k_i = k_{iR}+k_{iT})$
$P_{n_R n_T}(\underline{k}) = \Pr\{N_\Sigma$ is in state $k \mid$ number of class R(T)-jobs is n_R $(n_T)\}$
are given by |14|

$$P_{n_R n_T}(\underline{k}) = \gamma^{-1} \prod_{i=1}^{n} \frac{k_i!}{k_{iR}!\,k_{iT}!} \left(\frac{e_{iR}}{\mu_{iR}}\right)^{k_{iR}} \left(\frac{e_{iT}}{\mu_{iT}}\right)^{k_{iT}} \qquad (11)$$

where γ is the normalizing constant, and e_{iC} is the solution of the following traffic equation

$$e_{iC} = \sum_{j=1}^{n} e_{jC}\, p_{ji}(C)\,. \qquad (12)$$

Equation (11) holds if the ith unit is of Type 1 (FCFS discipline, single server with exponentially distributed and class independent service time), of Type 2 (round-robin processor-sharing) or of Type 4 (single server with preemptive LCFS discipline). For type 2 and 4 service time distributions must have a rational Laplace transform. We assume that no service facility is saturated.

The measures of interest are again device utilization and throughput. By

$$C^{i}_{n_R n_T} = \mu_{iC} \sum_{k_i>0} \frac{k_{iC}}{k_i} P_{n_R n_T}(\underline{k})\,,\ C = R \text{ or } T, \qquad (13)$$

we denote the σ_i-throughput of class C jobs; $\frac{k_{iC}}{k_i}$ is the fraction of the σ_i-capacity used by class C jobs. Thus, the systems's throughput is given by

$$R_{n_R n_T} = (1-p)\, R^1_{n_R n_T} . \tag{14}$$

Formula (11) yields for $\mu_{iT} = \tau$, $\mu_{iR} = \mu$, $i=1,2,\dots,n$:

$$R_{n_R O} = \mu(1-p)\,\frac{g(n_R)}{g(n_R+1)} \tag{15}$$

where

$$g(n_R) = \begin{cases} 1-p^{n_R} & , n=2 \\ 1-(n_R+1)p^{n_R} + n_R p^{n_R+1} & , n=3 \end{cases} \tag{16}$$

For $n_T > 0$ we obtain

$$P_{n_R n_T}(k) = \hat{\gamma}^{-1}\, p^{-k_{1R}} \prod_{i=1}^{n} \binom{k_i}{k_{iT}} \tag{17}$$

A numerical example:
Figs. 8, 9 show the (normalized) system throughput in the absence of testing and as the number of units and regular jobs is varied; $p_{11}(R) = 1-p$; $p_{12}(R) = p$. Tab. 3 gives numerical examples for n=3 and $n_T=1$ or 2. Test mode B is more effective since each test requests only a single unit and, hence, more jobs can be completed per unit time.

6. Conclusion

We have reported on an initial investigation into the performance of self-testing systems. In our model, a self-testing system is assumed to be a collection of units, such as people , microprocessors, etc., each of which is capable of applying tests to other units as well as being tested by them. The testing arrangement is represented by a directed graph. The routing of test- and normal jobs within the system is represented by queueing networks. We proposed a suitable measure which takes into account the diagnostic power as well as the regular processing power of self-testing systems. Because of lacking experimental information on self-testing systems, valididation of our models is not yet possible. However, the near future will see a variety of self-testing computer systems from which the necessary experimental information will be obtained.

List of symbols

$\lvert W \rvert$ number of elements in set W	$W-V = \{w \mid w \in W, w \notin V\}$
pr_i projection	$\bar{W}$ complement of W
$\lfloor a \rfloor$ largest integer not exceeding $a \in \mathbb{R}$	$\underline{1}$ row vector of 1's

No	A	$\varepsilon = \beta = 1$: $L_{AD}^{n_T}$	D_{2n_T}	$L_{AO}^{n_T}$	$\varepsilon = \beta = 2$: $L_{AD}^{n_T}$	D_{2n_T}	$L_{AO}^{n_T}$
1	1	.146	.103	.103	.177	.126	.126
2	.632	.103	.099	.065	.126	.119	.079
3	.512	.077	.090	.045	.107	.113	.064
Test mode A : $\delta = 1$, $L_{10} = .149^{+)}$, $n_T = 1$							
1	1	.148	.116	.116	.208	.147	.147
2	.632	.116	.110	.073	.147	.140	.093
3	.512	.098	.104	.059	.119	.113	.075
Test mode A: $\delta = 2$, $L_{10} = .149^{+)}$, $n = 1$							
1	1	.195	.200	.119	.206	.333	.099
2	.632	.141	.190	.076	.154	.316	.063
3	.512	.121	.180	.061	.134	.360	.051
Test mode B: $n_T = 1$, $L_{10} = .149$; $n_T = 2$							

Tab. 3 Numerical examples
Test mode A and B: $\mu = \tau = \lambda_i = \alpha = 1$, $p = \frac{3}{4}$
1) $\lambda_i = 0$, 2) $\rho_1 = 2\rho_2 = \rho_3 = 8$, 3) $\rho_i = 4$

+) L_{10} is given by $\tau = 0$ and appropriate changes of Fig. 7 (e.g., $p(2,2_E^1,8) = p$ and $p(2,2_E^1,6) = 0$)

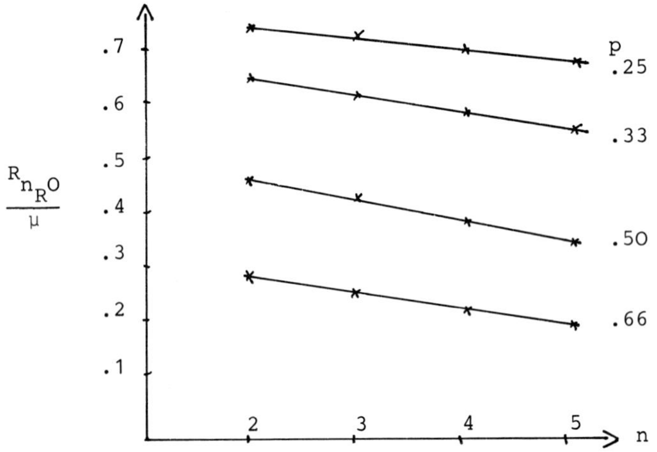

Fig. 8 System throughput, $n_R = 3$

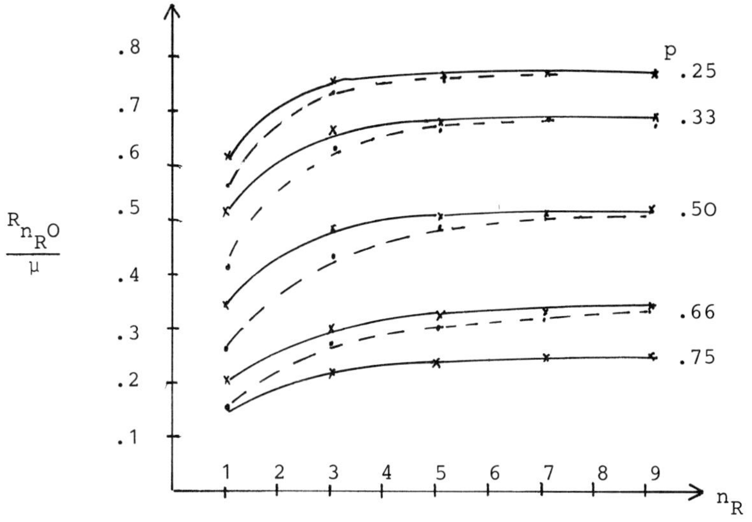

Fig. 9 System throughput, n = 2 ———, n = 3 -----

References

|1| von Neumann, J.; Probabilistic Logic and the Synthesis of Reliable Organisms from Unreliable Components, Automata Studies, Annals of Math. Study, 34, 1956

|2| von Neumann, J.; Theory of Self-Producing Automata, (ed. A.W. Burks), Univ. of Illinois Press, 1966

|3| Srini, V.P.; Fault Diagnosis of Microprocessor Systems, Computer, Jan. 1977, 60- 65, 1977

|4| Spetz, W.L.; Microprocessor Networks, Computer, July 1977, 64 - 70

|5| Preparata, R.P.; Metze, G.; Chien, R.T.; On the Connection Assignment Problem of Diagnosable Systems, IEEE Trans. on EC, Vol. EC 16, 848, 1967

|6| Hakimi, S.L.; Amin, A.T.; Characterization of Connection Assignment of Diagnosable Systems, IEEE Trans. on Comp., Jan. 1974, 86-87, 1974

|7| Cox, D.R.; Renewal Theory, Methuen, 1962

|8| Hammer, P.L.; Peled, U.N.; On Maximization of a Pseudo-Boolean Function, J. AMC 19, 265 - 282, 1972

|9| Maheshwari, S.; Hakimi, S.L.; On Models for Diagnosable Systems and Probabilistic Fault Diagnosis, IEEE Trans. on Comp., Vol. C-25, 228, 1976

|10| Chiu, W.; Dumont, D.; Wood, R.; Performance Analysis of a Multiprogrammed Computer System, IBM J. Res. Develop., May 1975, 263 - 271, 1975

|11| Dal Cin, M.; Performance Evaluation of Self-Diagnosing Multiprocessing Systems, Tübingen, 1977

|12| Dal Cin, M.; Analytic Models for Self-Testing Systems, Proceedings of the II. Symposium on OR, Aachen, Anton Hain Verlag, 1977

|13| König, D.; Jansen, U.; Eine Invarianzeigenschaft zufälliger Bedienungsprozesse mit positiven Geschwindigkeiten, Math. Nachr. Bd. 70, 321, 1976

|14| Baskett, F.; Chandy, K.M.; Muntz, R.R.; Palacios, F.G.; Open, Closed, and Mixed Networks of Queues with Different Classes of Customers, J. ACM, Vol. 22, 248, 1975

Bio–mathematics

Managing Editors: K. Krickeberg, S. A. Levin

Volume 1:
Mathematical Topics in Population Genetics
Edited by K. Kojima
1970. 55 figures. IX, 400 pages
ISBN 3-540-05054-X

"...It is far and away the most solid product I have ever seen labelled biomathematics."
American Scientist

Volume 2: E. Batschelet
Introduction to Mathematics for Life Scientists
2nd edition. 1975. 227 figures. XV, 643 pages
ISBN 3-540-07293-4

"A sincere attempt to relate basic mathematics to the needs of the student of life sciences."
Mathematics Teacher

M. Iosifescu, P. Tăutu
Stochastic Processes and Applications in Biology and Medicine

Volume 3
Part 1: **Theory**
1973. 331 pages.
ISBN 3-540-06270-X

Volume 4
Part 2: **Models**
1973. 337 pages
ISBN 3-540-06271-8

Distributions Rights for the Socialist Countries: Romlibri, Bucharest

"... the two-volume set, with its very extensive bibliography, is a survey of recent work as well as a textbook. It is highly recommended by the reviewer."
American Scientist

Volume 5: A. Jacquard
The Genetic Structure of Populations
Translated by B. Charlesworth, D. Charlesworth
1974. 92 figures. XVIII, 569 pages
ISBN 3-540-06329-3

"...should take its place as a major reference work.."
Science

Volume 6: D. Smith, N. Keyfitz
Mathematical Demography
Selected Papers
1977. 31 figures. XI, 515 pages
ISBN 3-540-07899-1

This collection of readings brings together the major historical contributions that form the base of current population mathematics tracing the development of the field from the early explorations of Graunt and Halley in the seventeenth century to Lotka and his successors in the twentieth. The volume includes 55 articles and excerpts with introductory histories and mathematical notes by the editors.

Volume 7: E. R. Lewis
Network Models in Population Biology
1977. 187 figures. XII, 402 pages
ISBN 3-540-08214-X

Directed toward biologists who are looking for an introduction to biologically motivated systems theory, this book provides a simple, heuristic approach to quantitative and theoretical population biology.

Springer-Verlag
Berlin
Heidelberg
New York